Implementation of Artificial Intelligence in Food Science, Food Quality, and Consumer Preference Assessment

Implementation of Artificial Intelligence in Food Science, Food Quality, and Consumer Preference Assessment

Editor

Sigfredo Fuentes

MDPI • Basel • Beijing • Wuhan • Barcelona • Belgrade • Manchester • Tokyo • Cluj • Tianjin

Editor
Sigfredo Fuentes
Agriculture and Food
The University of Melbourne
Parkville
Australia

Editorial Office
MDPI
St. Alban-Anlage 66
4052 Basel, Switzerland

This is a reprint of articles from the Special Issue published online in the open access journal *Foods* (ISSN 2304-8158) (available at: www.mdpi.com/journal/foods/special_issues/Implementation_Artificial_Intelligence_Food_Science_Food_Quality_Consumer_Preference_Assessment).

For citation purposes, cite each article independently as indicated on the article page online and as indicated below:

LastName, A.A.; LastName, B.B.; LastName, C.C. Article Title. *Journal Name* **Year**, *Volume Number*, Page Range.

ISBN 978-3-0365-4080-1 (Hbk)
ISBN 978-3-0365-4079-5 (PDF)

Contents

foods

MDPI

Editorial

Implementation of Artificial Intelligence in Food Science, Food Quality, and Consumer Preference Assessment

Sigfredo Fuentes 

Digital Agriculture, Food and Wine Research Group, School of Agriculture and Food, Faculty of Veterinary and Agricultural Sciences. The University of Melbourne, Parkville, VIC 3010, Australia; sfuentes@unimelb.edu.au; Tel.: +61-3-9035-9670

In recent years, new and emerging digital technologies applied to food science have been gaining attention and increased interest from researchers and the food/beverage industries. In particular, those digital technologies that can be used throughout the food value chain are accurate, easy to implement, affordable, and user-friendly. Hence, this Special Issue (SI) is dedicated to novel technology based on sensor technology and machine/deep learning modeling strategies to implement artificial intelligence (AI) into food and beverage production and for consumer assessment. This SI published quality papers from researchers in Australia, New Zealand, the United States, Spain, and Mexico, including food and beverage products such as grapes and wine [1], chocolate [2], honey [3], whiskey [4], avocado pulp [5], and a variety of other food products [6].

The analysis of big data, such as meteorological and vineyard management information using machine learning algorithms, has been used to target the prediction of aroma profiles for the Pinot Noir cultivar in Australia [1]. Wine aroma and chemometric profile prediction using readily available ancillary information could offer the viticulture and winemaking industries the advantage of characterizing wine regions and specific styles of wine production through vertical vintages. The accuracy of the regression models presented in this paper ($R > 0.94$) can be used to improve or maintain wine quality traits and styles for other wine regions using big data analysis. On the other hand, the quality analysis of chocolate has been based on a digital analysis using machine learning of chemical fingerprinting using near-infrared spectroscopy (NIR) [2]. This paper offered a non-destructive digital method to automatically assess physicochemical and sensory data to potentially achieve digital twins to assess chocolate quality traits more consistently, objectively, and affordably to the industry. The regression machine learning models developed also achieved high accuracy ($R > 0.93$). The classification of unfloral kinds of honey into botanical classes using the standard counting of pollen grains may be a daunting task. Research from Spain [3] proposed using a comparative analysis of a machine learning algorithm's performances to expedite this classification based on physicochemical parameters obtained from honey samples as inputs and honey classes based on botanical origins as targets. Eleven different ML algorithms were tested, with the penalized discriminant analysis (PDA) being the best performing one for overall accuracy. Interestingly, supervised vector machine (SVM) was the best performing algorithm, which may contradict the use of SVM for other applications published elsewhere. Sensory analysis is an area of research that is increasing the inclusion of digital technologies to assess the subconscious responses of consumers, which can offer a better understanding of the liking and appreciation of different cultures. However, one of the main bottlenecks can be found in the lexicon used when describing food and beverage products. A study from the USA used deep learning algorithms for the sensory descriptor of whiskey lexicon related to flavor characterization [4]. For this purpose, an interactive visual tool was implemented to tag samples of a descriptive lexicon from a database of whiskey reviews. The model proposed was able to identify descriptors with 99% accuracy. This research may facilitate lexicons for other food and beverage products that can also

Citation: Fuentes, S. Implementation of Artificial Intelligence in Food Science, Food Quality, and Consumer Preference Assessment. *Foods* **2022**, *11*, 1192. https://doi.org/10.3390/foods11091192

Received: 30 March 2022
Accepted: 19 April 2022
Published: 20 April 2022

target different cultural backgrounds and perceived terminology that can be equivalent to a more familiar one with specific consumers. Furthermore, the importance of sensory drivers for assessing food products such as avocado pulp of different cultivars has been investigated by a research team in Mexico using predictive modeling strategies [5]. Specific descriptive flavors, textural sensory drivers, instrumental stickiness, and color were found within the map modeling strategy implemented that can be useful for selecting avocado fruits to develop particular products with maximum acceptability by consumers. Finally, a research group from Mexico studied the digital assessment of consumers' physiological responses to sensory analysis of different products, including facial emotional recognition, galvanic skin response, and heart rate [6]. The integration of different sensors and analysis using machine learning algorithms targeted consumer acceptability with the best prediction, compared to the use of individual sensor technologies. The authors proposed using integrative biometric systems to completely predict the sensory responses using physiological responses alone to assess new food products.

Funding: This research received no external funding.

Conflicts of Interest: The author declares no conflict of interest.

References

1. Fuentes, S.; Tongson, E.; Torrico, D.D.; Gonzalez Viejo, C. Modeling pinot noir aroma profiles based on weather and water management information using machine learning algorithms: A vertical vintage analysis using artificial intelligence. *Foods* **2019**, *9*, 33. [CrossRef] [PubMed]
2. Gunaratne, T.M.; Gonzalez Viejo, C.; Gunaratne, N.M.; Torrico, D.D.; Dunshea, F.R.; Fuentes, S. Chocolate quality assessment based on chemical fingerprinting using near infra-red and machine learning modeling. *Foods* **2019**, *8*, 426. [CrossRef] [PubMed]
3. Mateo, F.; Tarazona, A.; Mateo, E.M. Comparative Study of Several Machine Learning Algorithms for Classification of Unifloral Honeys. *Foods* **2021**, *10*, 1543. [CrossRef] [PubMed]
4. Miller, C.; Hamilton, L.; Lahne, J. Sensory Descriptor Analysis of Whisky Lexicons through the Use of Deep Learning. *Foods* **2021**, *10*, 1633. [CrossRef] [PubMed]
5. Marín-Obispo, L.M.; Villarreal-Lara, R.; Rodríguez-Sánchez, D.G.; Follo-Martínez, D.; Espíndola Barquera, M.d.l.C.; Díaz de la Garza, R.I.; Hernández-Brenes, C. Insights into drivers of liking for avocado pulp (*persea americana*): Integration of descriptive variables and predictive modeling. *Foods* **2021**, *10*, 99. [CrossRef] [PubMed]
6. Álvarez-Pato, V.M.; Sánchez, C.N.; Domínguez-Soberanes, J.; Méndoza-Pérez, D.E.; Velázquez, R. A multisensor data fusion approach for predicting consumer acceptance of food products. *Foods* **2020**, *9*, 774. [CrossRef] [PubMed]

Article

Modeling Pinot Noir Aroma Profiles Based on Weather and Water Management Information Using Machine Learning Algorithms: A Vertical Vintage Analysis Using Artificial Intelligence

Sigfredo Fuentes [1,*], Eden Tongson [1], Damir D. Torrico [1,2] and Claudia Gonzalez Viejo [1]

[1] School of Agriculture and Food, Faculty of Veterinary and Agricultural Sciences, University of Melbourne, Melbourne, VIC 3010, Australia; eden.tongson@unimelb.edu.au (E.T.); damir.torrico@lincoln.ac.nz (D.D.T.); cgonzalez2@unimelb.edu.au (C.G.V.)

[2] Department of Wine, Food and Molecular Biosciences, Faculty of Agriculture and Life Sciences, Lincoln University, Lincoln 7647, New Zealand

* Correspondence: sfuentes@unimelb.edu.au; Tel.: +61-4245-04434

Received: 12 November 2019; Accepted: 27 December 2019; Published: 30 December 2019

Abstract: Wine aroma profiles are determinant for the specific style and quality characteristics of final wines. These are dependent on the seasonality, mainly weather conditions, such as solar exposure and temperatures and water management strategies from veraison to harvest. This paper presents machine learning modeling strategies using weather and water management information from a Pinot noir vineyard from 2008 to 2016 vintages as inputs and aroma profiles from wines from the same vintages assessed using gas chromatography and chemometric analyses of wines as targets. The results showed that artificial neural network (ANN) models rendered the high accuracy in the prediction of aroma profiles (Model 1; R = 0.99) and chemometric wine parameters (Model 2; R = 0.94) with no indication of overfitting. These models could offer powerful tools to winemakers to assess the aroma profiles of wines before winemaking, which could help adjust some techniques to maintain/increase the quality of wines or wine styles that are characteristic of specific vineyards or regions. These models can be modified for different cultivars and regions by including more data from vertical vintages to implement artificial intelligence in winemaking.

Keywords: wine quality; machine learning modeling; weather

---

1. Introduction

Wine quality traits are difficult to assess in a rapid and objective way in vineyards, especially before winemaking. Usually, quality assessments that are performed in the wine industry are related to the acidity and sugar content in berries (Brix or Baume) to assess maturity [1,2]. However, this assessment only gives information about the amount of alcohol and acidity in the final wine through fermentation. Hence, berry sugars/acidity do not provide useful information on any other important quality trait, such as the potential aroma profiles that could be obtained in the final wine.

Alcohol present in beverages has been found to have an effect on the perception of flavor and aromas, as it aids in the release of volatile aromatic compounds [3]. Furthermore, higher alcohol wines have been sometimes regarded as beneficial for the physicochemical expression of color and other quality traits that impact their sensory evaluation [4]. However, increasing the alcohol content in wines is a problem nowadays due to climate change, specifically global warming. Specifically, higher temperatures are compressing phenological stages, resulting in earlier harvest during hotter months around the globe [5–8]. This phenomenon produces a double global warming effect in grapevines, which can result in berry shrivel with the associated concentration of sugar in berries, and the degradation of color and aroma compounds, which impact the sensory aroma and flavor profile of final wines [7,9]. Recently, the assessment of mesocarp living tissue has been associated with quality traits for different grapevine cultivars for winemaking [10]. Berry cell death starts around 90 days after full bloom; it is a programmed cell death, which can be uncoupled from sugar accumulation and berry shrivel (both exacerbated by higher temperatures) and can determine the final quality of wines, aroma profile, and sensory appreciation [11,12]. Hence, there is a direct link between the seasonal weather characteristics, which are mainly temperature expressed in thermal time (degree days) accumulated over 10 °C and phenological stages occurrence and duration [13], berry cell death, wine quality, and aroma profiles [11,12]. Furthermore, these berry quality traits can be manipulated using different irrigation techniques, such as regulated deficit irrigation (RDI) [14–20] and partial rootzone drying (PRD) [21–25].

Some methods using proximal remote sensing within the near-infrared (NIR) light spectrum reflectivity have been developed to assess quality traits from berries in a non-destructive way. Some applications have been implemented to assess the sugar content in berries [26,27], berry pigments [28,29], phenolic compounds [30,31], and grape maturity in general [28,32–34]. However, since these techniques are still manual, they cannot account for the natural intra-bunch and vineyard spatial variability, requiring a huge number of measurements and modeling strategies to obtain meaningful results.

Other techniques have been developed thanks to recent advances in unmanned aerial vehicles and remote sensing techniques to assess grape maturity, which can take into account within-vineyard variability using high-resolution multispectral imagery analysis [35–37]. However, studies have been limited to a few flights per season, and the indirect assessment of berry quality and maturity may hamper results. Furthermore, associated costs for data acquisition, post-processing to obtain orthomosaics, data analysis for classification, and thematic map production are still costly, requiring in many countries licensed pilots and high data analysis power to obtain meaningful models.

This paper presents machine learning modeling strategies applying integrated vineyard weather and irrigation management parameters as inputs and the aroma profiles as targets obtained from a vertical wine library from a boutique vineyard. The results from this modeling strategy could offer an important tool to winemakers to assess the aroma profiles for future vintages before winemaking. The knowledge of potential aroma profiles of the final wine may allow making adjustments within the winemaking to maintain or increase quality traits in the final wine to maintain a specific wine style that is characteristic of the wine region or particular vineyard.

2. Materials and Methods

2.1. Study Area and Weather/Irrigation Management Data Acquisition

The study was conducted using weather and management data and wine samples from a vertical wine library belonging to a commercial vineyard located at an elevation of 540 m.a.s.l in the South of the Great Dividing Range of the Macedon Ranges in the sub-region of Romsey/Lancefield, Victoria in Australia. The vineyard is situated at a distance from the mitigating influence of the ocean (Figure 1), and the cultivars planted consist of 69% Pinot noir, 26% Chardonnay, and 5% Pinot gris, and use mostly the lyre training system. The study was conducted for vertical vintages from 2008 to 2016 of Pinot noir cultivars, and weather/irrigation management data were obtained from the same site for each season. Information such as (i) solar exposure from veraison to harvest (V-H), (ii) solar exposure from September to harvest (S-H), (iii) maximum January solar exposure (MJSE), (iv) degree days from S-H (DD-S-H), (v) maximum January temperature (MJT), (vi) mean maximum temperature from V-H (MeanMaxTV-H), and (vii) mean minimum temperature from V-H (MeanMinTV-H) was extracted from the Bureau of Meteorology (BoM). Furthermore, the water balance (WB) was calculated using the irrigation (I), rainfall (RF), and evapotranspiration (ET_c) data using the following Equation (1):

$$WB = I + RF(0.85) - ET_c \quad (1)$$

where WB = water balance; I = irrigation applied in megaliter (ML); RF = effective rainfall, considering 85% of the water is available to the plant, and ET_c = crop evapotranspiration calculated using the corresponding crop coefficient (K_c) for different phenological stages [14].

Figure 1. Aerial image of the study area obtained using an unmanned aerial vehicle (UAV) in the 2015–2016 growing season from a total area planted of 42 hectares.

2.2. Physicochemical Analysis

Wines from each vintage were analyzed in triplicates for the different physicochemical data measured in this study. A volume of 20 mL of each wine sample was poured in a 60 × 15 mm Greiner Bio-One Polystyrene Petri dish (item number 628102; Greiner Bio-One, Kremsmünster, Austria) and placed on a white uniform surface. Color in CIELab and RGB scales was measured using a NIX Pro color sensor (NIX Sensor Ltd. Hamilton, Ontario, Canada). The UV-Vis spectra from 380 to 780 nm were acquired with a Lighting Passport Pro portable spectrometer (Asensetek Incorporation, New Taipei City, Taiwan). To calculate color intensity, the absorbance of 420, 520, and 620 nm were summed, while for color hue, the absorbance from 420 nm was divided by the value from 520 nm. Fifty mL of each wine sample were used to determine liquid density (weight divided by volume), pH was determined using a pH-meter (QM-1670, DigiTech, Sandy, UT, USA), total dissolved solids (TDS) and electric conductivity (EC) were measured with a Yuelong YL-TDS2-A digital water quality tester (Zhengzhou Yuelong Electronic Technology Co., Ltd, Zhengzhou City, Henan Province, China), salt concentration was measured using a digital salt-meter (PAL-SALT Mohr, Atago Co., Ltd. Saitama, Japan), and alcohol content using an AlcolyzerWine M alcohol meter (Anton Paar GmbH, Graz, Austria).

2.3. Gas Chromatography–Mass Spectroscopy

A 5 mL sample of each wine replicate was poured into a 20 mL screw cap vial and sealed with an 18 mm magnetic screwcap with a polytetrafluoroethylene and silicone liner. These samples were analyzed with the method proposed by Gonzalez Viejo et al. [38] using a high-efficiency gas chromatograph with a mass selective detector 5977B (GC-MSD; Agilent Technologies, Inc., Santa Clara, CA, USA), coupled with a PAL3 autosampler system (CTC Analytics AG, Zwingen, Switzerland). The GC-MSD has a detection limit of 1.5 fg, and an HP-5MS column was attached (length: 30 m, inner diameter: 0.25 mm, film: 0.25 μ; Agilent Technologies, Inc., Santa Clara, CA, USA), while the flow rate was set to 1 mL min^{-1} of the carrier gas (Helium). Headspace with solid-phase microextraction (SPME) and a divinylbenzene–carboxen–polydimethylsiloxane grey fiber (1.1 mm; Agilent Technologies, Inc., Santa Clara, CA, USA) was used. Incubation time was set to 20 at 45 °C with a 5 min cycle and 1 min for fiber conditioning (170 °C). Furthermore, the extraction time was set to 40 min with agitation. Two blank samples were used, one at the start and one at the end to avoid any carryover effect. To identify the volatile compounds, the National Institute of Standards and Technology library (NIST; National Institute of Standards and Technology, Gaithersburg, MD, USA) was used. Only the compounds with ≥ 80% certainty were reported.

2.4. Statistical Analysis and Machine Learning Modeling

Data from weather, physicochemical, and aroma profile measurements were analyzed using a customized code written in Matlab® R2019a (Mathworks, Inc. Natick, MA, USA) to assess significant correlations ($p < 0.05$) between parameters were reported in a matrix. These data were also used to develop machine learning models based on artificial neural networks (ANN) using an automated code in Matlab® that tests 17 different training algorithms in a loop. The weather data related to (i) solar exposure V-H, (ii) solar exposure from S-H, (iii) MJSE, (iv) DD-S-H, (v) MJT, (vi) MeanMaxTV-H, (vii) MeanMinTV-H, and (viii) water balance were used as inputs for machine learning purposes. Two models were developed using these inputs to predict (i) the peak area of nine volatile aromatic compounds measured using the GC-MSD (Model 1) and (ii) 14 physicochemical measurements (Model 2). Both models were developed using normalized data (inputs and targets) from −1 to 1, and with a random data division with 60% of the samples used for training with a Levenberg–Marquardt algorithm, 20% for validation with a mean squared error performance algorithm, and 20% for testing with a default derivative function. The number of neurons was defined by performing a trimming exercise with three, five, seven, and 10 neurons, with 10 neurons giving the best models that contribute to the absence of overfitting. The models consisted of a two-layer feedforward network with a tan-sigmoid function in the hidden layer and a linear transfer function in the output layer (Figure 2).

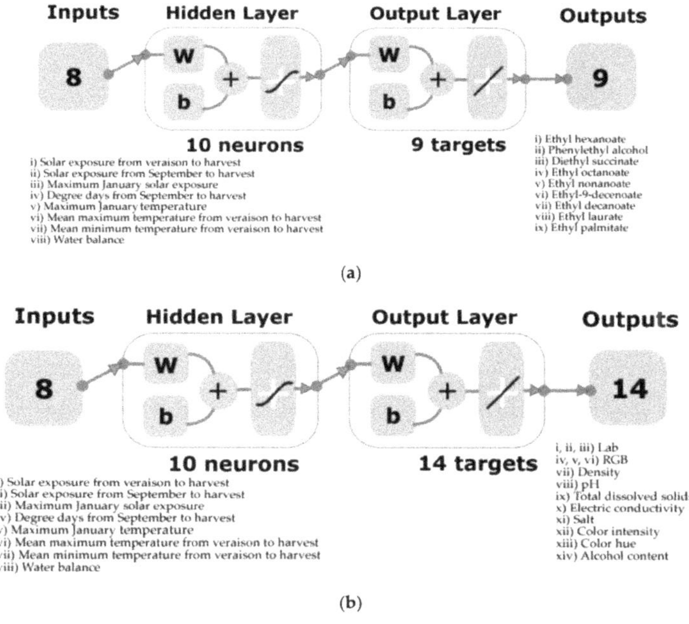

Figure 2. Artificial neural network model diagrams showing the inputs and target/outputs of (**a**) Model 1 to predict the aroma profile based on the peak area of volatile aromatic compounds, and (**b**) the physicochemical data of Pinot noir wines.

3. Results

Table 1 shows the mean values of the weather data for the vintages with contrasting water balance data (2011–2014). It can be observed that 2011 was the wettest season with the lowest solar exposure and mean temperatures (MeanMaxTV-H and MeanMinTV-H), while 2013 was the driest with the highest MJSE and solar exposure. Vintages 2012 and 2014 presented values in the mid-range.

Table 1. Mean values of weather data only for the contrasting vintages based on water balance.

Year	Solar Exposure (V-H; MJ $m^{2\ -1}$)	Solar Exposure (S-H; MJ $m^{2\ -1}$)	MJSE (MJ $m^{2\ -1}$)	DD-S-H (days)	MJT (°C)	MeanMaxT V-H (°C)	Mean MinTV-H (°C)	Water Balance (mm)
2011	15.6	19.1	24.6	1066.8	18.6	19.7	9.44	673.7
2012	17.9	20.2	26.3	1147.3	19.4	22.6	10.75	255.9
2013	21.8	21.8	28.9	1234.2	19.8	26.1	12.05	−117.5
2014	19.0	20.0	27.6	1223.7	20.3	25.8	11.31	−61.9

Abbreviations: V-H = veraison to harvest, S-H = September to harvest, MJSE = maximum January solar exposure, DD = degree days, MJT = maximum January temperature, MaxTV-H = maximum temperature veraison to harvest, MinTV-H minimum temperature veraison to harvest.

Table 2 shows the nine volatile compounds identified in all the wine samples tested and the aromas associated with them. It can be observed from this table that most of the aromas are related to fruity scents, especially apple, with two specific compounds (phenylethyl alcohol and ethyl laurate) with floral and one (ethyl palmitate) with milky or creamy notes.

Table 2. Volatile compounds identified using gas chromatography–mass spectroscopy and their associated aromas.

Volatile Compound	Aroma *
Ethyl hexanoate	Apple/Green banana/Pineapple
Phenylethyl alcohol	Rose/Bread/Honey
Diethyl succinate	Cooked apple
Ethyl octanoate	Apple/Banana/Pineapple
Ethyl nonanoate	Cognac/Apple/Winey/Nutty
Ethyl-9-decenoate	Fruity/Fatty/Roses
Ethyl decanoate	Waxy/Apple/Grape
Ethyl laurate	Floral/Soapy/Sweet
Ethyl palmitate	Waxy/Fruity/Creamy/Milky

* The association between the volatile compounds and aromas were obtained from The Good Scents Company [39], Genovese et al. [40], Arcari et al. [41], and Gonzalez Viejo et al. [38].

Figure 3 shows the significant ($p < 0.05$) correlations between the weather information, the aromas, and physicochemical data. It can be observed that the solar exposure from September to harvest was positively correlated with diethyl succinate ($r = 0.90$), while the degree days from September to harvest was negatively correlated with ethyl-9-decenoate ($r = 0.88$). The MJT had a positive correlation with phenylethyl alcohol ($r = 0.82$) and "b" ($r = 0.88$), and a negative correlation with "B". The MeanMaxTV-H was negatively correlated with ethyl-9-decenoate ($r = -0.93$) and color intensity ($r = -0.90$), as well as positively correlated with color hue ($r = 0.92$) and "a" ($r = 0.84$). On the other hand, the MeanMinTV-H had a negative correlation with ethyl hexanoate ($r = -0.93$), TDS ($r = -0.90$), and EC ($r = -0.90$). Water balance was positively correlated with ethyl-9-decenoate ($r = 0.93$) and color intensity ($r = 0.90$), and negatively correlated with color hue ($r = -0.95$) and "a" ($r = -0.86$). Mean values of the aromatic volatile compounds and physicochemical data are shown as supplementary material in Table S1.

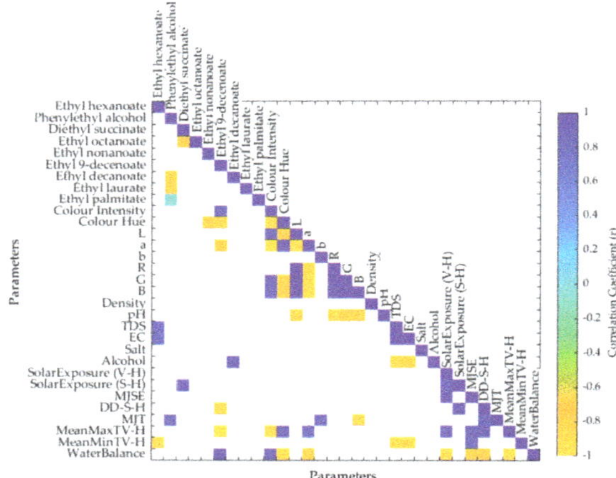

Figure 3. Matrix showing only the significant correlations ($p < 0.05$) between the weather and physicochemical data and volatile aromatic compounds of Pinot noir wines of vintages from 2008 to 2016. Abbreviations: TDS = total dissolved solids, EC = electric conductivity, V-H = veraison to harvest, S-H = September to harvest, MJSE = maximum January solar exposure, DD = degree days, MJT = maximum January temperature, MaxTV-H = maximum temperature veraison to harvest, MinTV-H minimum temperature veraison to harvest.

In Table 3, the statistical results from the ANN models are shown. Model 1 had an overall high correlation coefficient ($r = 0.99$) with similar results for all stages (training, validation, and testing; $r > 0.97$) to predict the peak area of nine volatile aromatic compounds (Table 2). From the performance, it can be observed that both validation and testing mean square error (MSE) values were the same (MSE = 0.03), and the training had a lower result (MSE = 0.003), which contributes to the absence of overfitting of the model. Furthermore, the slope (b) for all stages and the overall model was close to the unity ($b = 0.97$). On the other hand, Model 2 had an overall correlation $r = 0.94$ to predict 14 physicochemical parameters (Figure 2b). The slopes from the models of the three stages were high enough ($b > 0.83$) with an overall model $b = 0.90$. Similar to Model 1, the performance of the training stage from Model 2 was lower (MSE = 0.02) than the validation and testing stages, with the last two presenting similar results (MSE = 0.05 and MSE = 0.06; respectively).

Table 3. Statistics from the artificial neural network models to predict the aroma profile based on the peak area of volatile aromatic compounds (Model 1) and the physicochemical data (Model 2) from Pinot noir wines.

Stage	Samples	Observations	*R*	Slope (b)	Performance (MSE)
			Model 1		
Training	40	360	0.99	0.98	0.003
Validation	13	117	0.97	0.98	0.03
Testing	13	117	0.97	0.92	0.03
Overall	**66**	**594**	**0.99**	**0.97**	/
			Model 2		
Training	40	560	0.96	0.91	0.02
Validation	13	182	0.93	0.83	0.05
Testing	13	182	0.90	0.94	0.06
Overall	**66**	**924**	**0.94**	**0.90**	/

Abbreviations: R = correlation coefficient and MSE = mean square error.

Figure 4a shows the overall Model 1 to predict the aroma profile based on the peak area of volatile aromatic compounds of Pinot noir wines. From the 95% confidence bounds, only 1.01% of outliers (six out of 594) were found. On the other hand, Figure 4b depicts the overall Model 2 to predict the physicochemical data of the wines. Regarding the 95% prediction bounds, the model presented 3.25% (30 out of 924) of outliers. For both models, several retraining attempts were performed, obtaining similar results to those presented in Table 3 and Figure 4. When feeding these models with new data, the outputs values are given normalized from −1 to 1; however, the reverse function for normalization in Matlab® R2019a (Mathworks Inc., Natick, MA, USA) provides the actual values in the corresponding units.

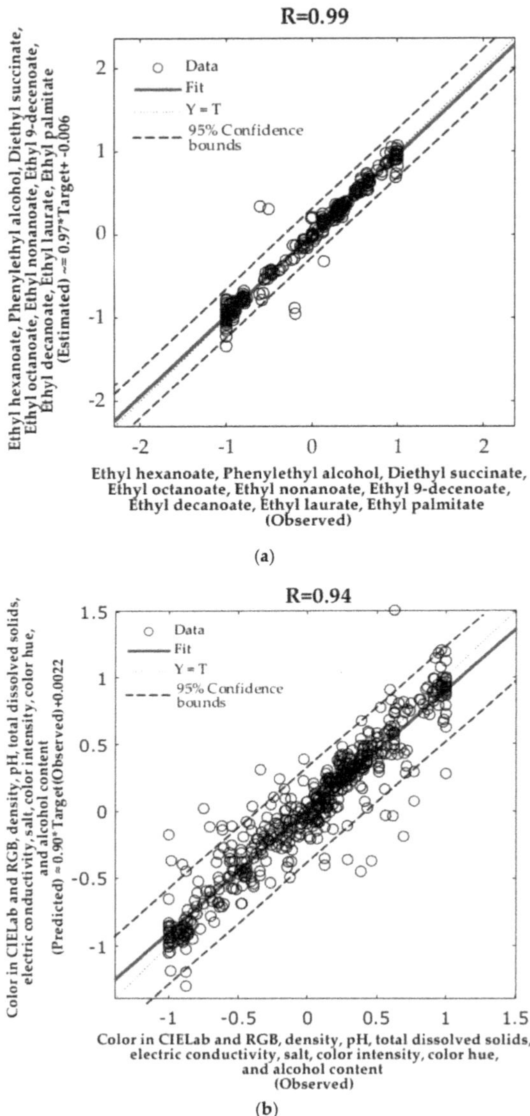

Figure 4. Overall artificial neural network models to predict (**a**) the aroma profile (Model 1) and (**b**) the physicochemical parameters of Pinot noir wines (Model 2), both using the weather data as inputs (Figure 2). The models show the observed (x-axis) and predicted (y-axis) data as well as the 95% confidence bounds.

4. Discussion

The physicochemical parameters assessed in this study have been associated with wine quality by other authors. Aromas and color-related parameters are some of the factors that have been the most associated with wine quality [42,43]. Sáenz-Navajas et al. [44] found that there is a relationship between red wine color and the quality perception from consumers and concluded that darker wines with higher red and lower yellow values were rated as higher quality. Jackson et al. [42] reported a significant and positive correlation between both pH and color and overall wine quality. The importance of TDS, EC, and salt measurements rely on the fact that these are an approach to minerals content [45], which are important in wine quality, as the minerals present in wine have been related to those present in the soil, and these have been associated with the wine's nutritional composition and safety [46].

There was a significant variability within the vintages and the particular region in Victoria analyzed in this study. The extremes can be considered for low-quality wines produced in the 2010–2011 vintage due to heavy rains before harvest, which negatively affects the quality traits in berries and wine [47,48]; this low-quality assessment was obtained from anecdotal information from points received in those particular years and the sensory analysis conducted by the vineyard studied. On the contrary, dry seasons were found for example in 2013–2014 and 2014–2015, with increased berry quality traits that were passed to the respective wines. The latter were mainly due to some control of the water received by plants from irrigation and water deficits. These differences contribute to the robustness of the machine learning models found, which presented no indication of overfitting with high precision in the prediction of the peak area of volatile aromatic compounds (Model 1) and physicochemical wine characteristics (Model 2).

The effects of solar exposure and canopy architecture (which is dependent on water balance) on the aroma profiles of wines have been previously reported, and they are consistent with the data presented in Figure 3. Specifically, these effects manifest through the influence of the microclimate within bunches [49], phenolic compounds [50,51], and the flavonol profile [52]. Due to the direct effect of bunch exposure to radiation in the aroma profiles obtained in wines, researchers have investigated the effect of defoliation as a management strategy to increase berry quality and aroma traits, which depends on the cultivar, timing of defoliation, and climatic region [53–60]. These researches demonstrate the importance of fruit exposure to solar radiation and microclimate conditions that are favorable to the development of berry quality traits.

As previously mentioned, seasonal temperatures not only influence the occurrence and length of different phenological stages in grapevines, such as budbreak, flowering, berry set, pea size, veraison, and harvest, but also the chemical and aroma composition of berries. Of critical importance is the influence of weather parameters, such as temperature [61–64], and water availability from veraison onwards in red cultivars, which is determinant to the final wine quality and aroma profiles. Several studies have focused on the pre and post veraison phenological stages for irrigation treatments to increase berry and wine quality traits, especially in red cultivars [65–69].

For machine learning modeling, it has been demonstrated that the implementation of important parameters as inputs that directly influence the targets proposed render more robust models in contrast to the usage of raw data. Based on calculated parameters rather than raw data inputs, there are recent studies implementing machine learning to assess beer quality [70–72], interpret remote sensing data for plant water status assessment in vineyards [73], chocolate quality assessment by consumers using NIR [74], and aroma profiles in cocoa trees based on canopy architecture parameters [75]. In this study, relevant parameters from weather conditions, management strategies, and physicochemical parameters of wines were obtained and considered as inputs in the machine learning modeling, which can explain the high accuracy obtained for the predictions of Models 1 and 2 without signs of overfitting.

The use of ANN for modeling has the advantage of being able to use multiple targets, which makes the models more efficient. This is due to the easiness of feeding only one model to obtain all the output data instead of having to add the new inputs to several single-target models. Several studies related to food and agriculture have used this type of machine learning algorithms with high performance and accuracy [38,71,72,75–77].

The technique proposed considers the readily available weather information from vintages close to the vineyards and a vertical vintage library, which most wineries can obtain easily. The models developed assume that the vineyard management is consistent throughout the seasons, including the winemaking techniques and yeast used. The implementation of these models to other cultivars, environments, and regions will need the incorporation of further site-specific data as inputs and wine chemical and aroma profile analysis from available and contrasting vintages. The latter benefit from the learning aspect of the models proposed, which does not require a full development of new analyses for different regions.

5. Conclusions

Artificial intelligence techniques can be implemented in the wine industry from readily available weather and management practices data to assess quality traits in final wines. Modeling strategies using artificial neural networks developed for particular regions can be implemented for other cultivars, environments, and regions by including extreme values from their respective vintages. High accuracy models to determine the aroma profile of wines before the winemaking process can offer a powerful tool to growers and winemakers for the decision making in the vinification process to maintain or increase wine quality and styles. Further research is required to adapt these techniques to canopy management strategies and within-season modeling that can be implemented in real-time within the season to manipulate the final wine and aroma profiles to specific targets using management strategies, such as canopy, fertilization, and irrigation management.

Supplementary Materials: The following are available online at http://www.mdpi.com/2304-8158/9/1/33/s1, Table S1: Means and standard error (SE) of the volatile aromatic compounds and physicochemical parameters of the wined from each vintage.

Author Contributions: Conceptualization, S.F.; Data curation, S.F., E.T. and C.G.V.; Formal analysis, S.F. and C.G.V.; Investigation, S.F. and C.G.V.; Methodology, D.D.T. and C.G.V.; Project administration, S.F.; Validation, S.F. and C.G.V.; Visualization, S.F. and C.G.V.; Writing—original draft, S.F. and C.G.V.; Writing—review & editing, E.T. and D.D.T. All authors have read and agreed to the published version of the manuscript.

Funding: This research received no external funding.

Acknowledgments: The authors acknowledge contributions of Xiaoyi Wang and Pangzhen Zhang for preliminary data handling.

Conflicts of Interest: The authors declare no conflict of interest.

References

1. Coombe, B.G.; Dundon, R.J.; Short, A.W. Indices of sugar—Acidity as ripeness criteria for winegrapes. *J. Sci. Food Agric.* **1980**, *31*, 495–502.
2. Suklje, K.; Blackman, J.; Deloire, A.; Schmidtke, L.; Antalick, G.; Meeks, C. Grapes to wine: The nexus between berry ripening, composition and wine style. In Proceedings of the X International Symposium on Grapevine Physiology and Biotechnology 1188, Verona, Italy, 13–18 June 2016; pp. 43–50.
3. Ickes, C.M.; Cadwallader, K.R. Effects of ethanol on flavor perception in alcoholic beverages. *Chemosens. Percept.* **2017**, *10*, 119–134.
4. Sherman, E.; Greenwood, D.R.; Villas-Boâs, S.G.; Heymann, H.; Harbertson, J.F. Impact of grape maturity and ethanol concentration on sensory properties of Washington State Merlot wines. *Am. J. Enol. Vitic.* **2017**, *68*, 344–356.
5. Webb, L.; Whetton, P.; Bhend, J.; Darbyshire, R.; Briggs, P.; Barlow, E. Earlier wine-grape ripening driven by climatic warming and drying and management practices. *Nat. Clim. Chang.* **2012**, *2*, 259.
6. Anderson, K.; Findlay, C.; Fuentes, S.; Tyerman, S. Viticulture, wine and climate change. In *Garnaut Climate Change Review*; Cambridge University Press: Cambridge, UK, 2008; pp. 1–16.
7. Webb, L.; Whetton, P.; Barlow, E. Modelled impact of future climate change on the phenology of winegrapes in Australia. *Aust. J. Grape Wine Res.* **2007**, *13*, 165–175.
8. Webb, L.; Whetton, P.; Barlow, E. Observed trends in winegrape maturity in Australia. *Glob. Chang. Biol.* **2011**, *17*, 2707–2719.
9. Krasnow, M.; Weis, N.; Smith, R.J.; Benz, M.J.; Matthews, M.; Shackel, K. Inception, progression, and compositional consequences of a berry shrivel disorder. *Am. J. Enol. Vitic.* **2009**, *60*, 24–34.
10. Fuentes, S.; Sullivan, W.; Tilbrook, J.; Tyerman, S. A novel analysis of grapevine berry tissue demonstrates a variety-Dependent correlation between tissue vitality and berry shrivel. *Aust. J. Grape Wine Res.* **2010**, *16*, 327–336. [CrossRef]

11. Bonada, M.; Sadras, V.; Moran, M.; Fuentes, S. Elevated temperature and water stress accelerate mesocarp cell death and shrivelling, and decouple sensory traits in Shiraz berries. *Irrig. Sci.* **2013**, *31*, 1317–1331. [CrossRef]
12. Bonada, M.; Sadras, V.O.; Fuentes, S. Effect of elevated temperature on the onset and rate of mesocarp cell death in berries of Shiraz and Chardonnay and its relationship with berry shrivel. *Aust. J. Grape Wine Res.* **2013**, *19*, 87–94. [CrossRef]
13. Molitor, D.; Junk, J.; Evers, D.; Hoffmann, L.; Beyer, M. A high-resolution cumulative degree day-based model to simulate phenological development of grapevine. *Am. J. Enol. Vitic.* **2014**, *65*, 72–80.
14. Acevedo-Opazo, C.; Ortega-Farias, S.; Fuentes, S. Effects of grapevine (Vitis vinifera L.) water status on water consumption, vegetative growth and grape quality: An irrigation scheduling application to achieve regulated deficit irrigation. *Agric. Water Manag.* **2010**, *97*, 956–964.
15. Santesteban, L.; Miranda, C.; Royo, J. Regulated deficit irrigation effects on growth, yield, grape quality and individual anthocyanin composition in Vitis vinifera L. cv.'Tempranillo'. *Agric. Water Manag.* **2011**, *98*, 1171–1179.
16. Greven, M.; Green, S.; Neal, S.; Clothier, B.; Neal, M.; Dryden, G.; Davidson, P. Regulated Deficit Irrigation (RDI) to save water and improve Sauvignon Blanc quality? *Water Sci. Technol.* **2005**, *51*, 9–17.
17. Casassa, L.; Keller, M.; Harbertson, J. Regulated deficit irrigation alters anthocyanins, tannins and sensory properties of Cabernet Sauvignon grapes and wines. *Molecules* **2015**, *20*, 7820–7844.
18. Permanhani, M.; Costa, J.M.; Conceição, M.; De Souza, R.; Vasconcellos, M.; Chaves, M. Deficit irrigation in table grape: Eco-physiological basis and potential use to save water and improve quality. *Theor. Exp. Plant Physiol.* **2016**, *28*, 85–108.
19. Romero, P.; Gil-Muñoz, R.; del Amor, F.M.; Valdés, E.; Fernández, J.I.; Martinez-Cutillas, A. Regulated deficit irrigation based upon optimum water status improves phenolic composition in Monastrell grapes and wines. *Agric. Water Manag.* **2013**, *121*, 85–101.
20. Ju, Y.-l.; Liu, M.; Tu, T.-y.; Zhao, X.-f.; Yue, X.-f.; Zhang, J.-x.; Fang, Y.-l.; Meng, J.-f. Effect of regulated deficit irrigation on fatty acids and their derived volatiles in 'Cabernet Sauvignon'grapes and wines of Ningxia, China. *Food Chem.* **2018**, *245*, 667–675.
21. Dry, P.; Loveys, B. Factors influencing grapevine vigour and the potential for control with partial rootzone drying. *Aust. J. Grape Wine Res.* **1998**, *4*, 140–148.
22. dos Santos, T.P.; Lopes, C.M.; Rodrigues, M.L.; de Souza, C.R.; Maroco, J.P.; Pereira, J.S.; Silva, J.R.; Chaves, M.M. Partial rootzone drying: Effects on growth and fruit quality of field-grown grapevines (Vitis vinifera). *Funct. Plant Biol.* **2003**, *30*, 663–671.
23. Bindon, K.; Dry, P.; Loveys, B. Influence of partial rootzone drying on the composition and accumulation of anthocyanins in grape berries (Vitis vinifera cv. Cabernet Sauvignon). *Aust. J. Grape Wine Res.* **2008**, *14*, 91–103.
24. Gil, P.; Lobos, P.; Duran, K.; Olguin, J.; Cea, D.; Schaffer, B. Partial root-zone drying irrigation, shading, or mulching effects on water savings, productivity and quality of 'Syrah'grapevines. *Sci. Hortic.* **2018**, *240*, 478–483.
25. Gotur, M.; Sharma, D.; Joshi, C.; Rajan, R. Partial root-zone drying technique in fruit crops: A review paper. *IJCS* **2018**, *6*, 900–903.
26. Urraca, R.; Sanz-Garcia, A.; Tardaguila, J.; Diago, M.P. Estimation of total soluble solids in grape berries using a hand-held NIR spectrometer under field conditions. *J. Sci. Food Agric.* **2016**, *96*, 3007–3016.
27. Wu, G.; Huang, L.; He, Y. Research on the sugar content measurement of grape and berries by using Vis/NIR spectroscopy technique. *Guang Pu Xue Yu Guang Pu Fen Xi* **2008**, *28*, 2090–2093.
28. Ribera-Fonseca, A.; Noferini, M.; Jorquera-Fontena, E.; Rombolà, A.D. Assessment of technological maturity parameters and anthocyanins in berries of cv. Sangiovese (Vitis vinifera L.) by a portable vis/NIR device. *Sci. Hortic.* **2016**, *209*, 229–235.
29. Chen, S.; Zhang, F.; Ning, J.; Liu, X.; Zhang, Z.; Yang, S. Predicting the anthocyanin content of wine grapes by NIR hyperspectral imaging. *Food Chem.* **2015**, *172*, 788–793.
30. Ferrer-Gallego, R.; Hernández-Hierro, J.M.; Rivas-Gonzalo, J.C.; Escribano-Bailón, M.T. Determination of phenolic compounds of grape skins during ripening by NIR spectroscopy. *LWT Food Sci. Technol.* **2011**, *44*, 847–853.

31. Gajdoš Kljusurić, J.; Mihalev, K.; Bečić, I.; Polović, I.; Georgieva, M.; Djaković, S.; Kurtanjek, Ž. Near-infrared spectroscopic analysis of total phenolic content and antioxidant activity of berry fruits. *Food Technol. Biotechnol.* **2016**, *54*, 236–242.
32. Herrera, J.; Guesalaga, A.; Agosin, E. Shortwave–near infrared spectroscopy for non-destructive determination of maturity of wine grapes. *Meas. Sci. Technol.* **2003**, *14*, 689.
33. Larraín, M.; Guesalaga, A.R.; Agosín, E. A multipurpose portable instrument for determining ripeness in wine grapes using NIR spectroscopy. *IEEE Trans. Instrum. Meas.* **2008**, *57*, 294–302.
34. Segade, S.R.; Giacosa, S.; Gerbi, V.; Rolle, L. Grape Maturity and Selection: Automatic Grape Selection. In *Red Wine Technology*; Elsevier: Amsterdam, The Netherlands, 2019; pp. 1–16.
35. Soubry, I.; Patias, P.; Tsioukas, V. Monitoring Vineyards with UAV and Multi-sensors for the assessment of Water Stress and Grape Maturity. *J. Unmanned Veh. Syst.* **2017**, *5*, 37–50.
36. Matese, A.; Di Gennaro, S.; Miranda, C.; Berton, A.; Santesteban, L. Evaluation of spectral-based and canopy-based vegetation indices from UAV and Sentinel 2 images to assess spatial variability and ground vine parameters. *Adv. Anim. Biosci.* **2017**, *8*, 817–822.
37. Matese, A.; Di Gennaro, S. Practical applications of a multisensor uav platform based on multispectral, thermal and rgb high resolution images in precision viticulture. *Agriculture* **2018**, *8*, 116.
38. Gonzalez Viejo, C.; Fuentes, S.; Torrico, D.D.; Godbole, A.; Dunshea, F.R. Chemical characterization of aromas in beer and their effect on consumers liking. *Food Chem.* **2019**, *293*, 479–485.
39. The Good Scents Company. The Good Scents Company Information System. Available online: http://www.thegoodscentscompany.com/data/rw1038291.html (accessed on 3 September 2019).
40. Genovese, A.; Piombino, P.; Gambuti, A.; Moio, L. Simulation of retronasal aroma of white and red wine in a model mouth system. Investigating the influence of saliva on volatile compound concentrations. *Food Chem.* **2009**, *114*, 100–107.
41. Arcari, S.G.; Caliari, V.; Sganzerla, M.; Godoy, H.T. Volatile composition of Merlot red wine and its contribution to the aroma: Optimization and validation of analytical method. *Talanta* **2017**, *174*, 752–766.
42. Jackson, M.G.; Timberlake, C.F.; Bridle, P.; Vallis, L. Red wine quality: Correlations between colour, aroma and flavour and pigment and other parameters of young Beaujolais. *J. Sci. Food Agric.* **1978**, *29*, 715–727.
43. Rankine, B.; Fornachon, J.; Boehm, E.; Cellier, K. Influence of grape variety, climate and soil on grape composition and on the composition and quality of table wines. *VITIS J. Grapevine Res.* **2017**, *10*, 33.
44. Sáenz-Navajas, M.-P.; Echavarri, F.; Ferreira, V.; Fernández-Zurbano, P. Pigment composition and color parameters of commercial Spanish red wine samples: Linkage to quality perception. *Eur. Food Res. Technol.* **2011**, *232*, 877–887.
45. Tariq, M.; Ali, M.; Shah, Z. Characteristics of industrial effluents and their possible impacts on quality of underground water. *Soil Environ.* **2006**, *25*, 64–69.
46. Karataş, D.; Aydin, I.; Karataş, H. Elemental composition of red wines in Southeast Turkey. *Czech J. Food Sci.* **2015**, *33*, 228–236.
47. Balint, G.; Reynolds, A.G. Irrigation level and time of imposition impact vine physiology, yield components, fruit composition and wine quality of Ontario Chardonnay. *Sci. Hortic.* **2017**, *214*, 252–272.
48. Romero, P.; García, J.G.; Fernández-Fernández, J.I.; Muñoz, R.G.; del Amor Saavedra, F.; Martínez-Cutillas, A. Improving berry and wine quality attributes and vineyard economic efficiency by long-term deficit irrigation practices under semiarid conditions. *Sci. Hortic.* **2016**, *203*, 69–85.
49. Martin, D.; Grose, C.; Fedrizzi, B.; Stuart, L.; Albright, A.; McLachlan, A. Grape cluster microclimate influences the aroma composition of Sauvignon blanc wine. *Food Chem.* **2016**, *210*, 640–647. [CrossRef]
50. Brillante, L.; Martínez-Lüscher, J.; Kurtural, S.K. Applied water and mechanical canopy management affect berry and wine phenolic and aroma composition of grapevine (Vitis vinifera L., cv. Syrah) in Central California. *Sci. Hortic.* **2018**, *227*, 261–271. [CrossRef]
51. Song, J.; Smart, R.; Wang, H.; Dambergs, B.; Sparrow, A.; Qian, M.C. Effect of grape bunch sunlight exposure and UV radiation on phenolics and volatile composition of Vitis vinifera L. cv. Pinot noir wine. *Food Chem.* **2015**, *173*, 424–431. [CrossRef]
52. Martínez-Lüscher, J.; Brillante, L.; Kurtural, S.K. Flavonol Profile Is a Reliable Indicator to Assess Canopy Architecture and the Exposure of Red Wine Grapes to Solar Radiation. *Front. Plant Sci.* **2019**, *10*, 10. [CrossRef]

53. Radovanović, V.; Stefanović, D.; Radovanović, B. Influence of selective removal of grapevine leaves on quality of red wine. *J. Process. Energy Agric.* **2015**, *19*, 215–218.
54. Verdenal, T.; Zufferey, V.; Dienes-Nagy, A.; Bourdin, G.; Gindro, K.; Viret, O.; Spring, J.-L. Timing and Intensity of Grapevine Defoliation: An Extensive Overview on Five Cultivars in Switzerland. *Am. J. Enol. Vitic.* **2019**, *70*, 427–434.
55. Vargas, S.; Cazorla, M.; Bordeu, E.; Casaubon, G.; González, Á. Evaluation of leaf removal strategies and cluster radiation protection on grape and wine quality of Vitis vinifera L'Cabernet Sauvignon. In Proceedings of the X International Symposium on Grapevine Physiology and Biotechnology 1188, Verona, Italy, 13–18 June 2016; pp. 97–104.
56. Pessenti, I.L.; Ayub, R.A.; Botelho, R.V. Defoliation, application of S-ABA and vegetal extracts on the quality of grape and wine Malbec cultivar. *Rev. Bras. Frutic.* **2019**, *41*. [CrossRef]
57. Wang, Y.; He, L.; Pan, Q.; Duan, C.; Wang, J. Effects of Basal Defoliation on Wine Aromas: A Meta-Analysis. *Molecules* **2018**, *23*, 779.
58. Scafidi, P.; Barbagallo, M.; Pisciotta, A.; Mazza, M.; Downey, M. Defoliation of two-wire vertical trellis: Effect on grape quality. *N. Z. J. Crop Hortic. Sci.* **2018**, *46*, 18–38.
59. Peña-Olmos, J.E.; Casierra-Posada, F. Fruit quality and production of Vitis vinifera L. Chardonnay affected by partial defoliation in tropical highlands. *Rev. Fac. Nac. De Agron. Medellín* **2015**, *68*, 7581–7588.
60. Baiano, A.; De Gianni, A.; Previtali, M.A.; Del Nobile, M.A.; Novello, V.; de Palma, L. Effects of defoliation on quality attributes of Nero di Troia (Vitis vinifera L.) grape and wine. *Food Res. Int.* **2015**, *75*, 260–269.
61. Jackson, D.; Lombard, P. Environmental and management practices affecting grape composition and wine quality-A review. *Am. J. Enol. Vitic.* **1993**, *44*, 409–430.
62. Drappier, J.; Thibon, C.; Rabot, A.; Geny-Denis, L. Relationship between wine composition and temperature: Impact on Bordeaux wine typicity in the context of global warming. *Crit. Rev. Food Sci. Nutr.* **2019**, *59*, 14–30.
63. Kliewer, W.M.; Torres, R.E. Effect of controlled day and night temperatures on grape coloration. *Am. J. Enol. Vitic.* **1972**, *23*, 71–77.
64. Teixeira, A.; Eiras-Dias, J.; Castellarin, S.D.; Gerós, H. Berry phenolics of grapevine under challenging environments. *Int. J. Mol. Sci.* **2013**, *14*, 18711–18739.
65. Romero, P.; Fernández, J.; Botía, P. Interannual climatic variability effects on yield, berry and wine quality indices in long-term deficit irrigated grapevines, determined by multivariate analysis. *Int. J. Wine Res.* **2016**, *8*, 3–17.
66. Intrigliolo, D.; Lizama, V.; García-Esparza, M.; Abrisqueta, I.; Álvarez, I. Effects of post-veraison irrigation regime on Cabernet Sauvignon grapevines in Valencia, Spain: Yield and grape composition. *Agric. Water Manag.* **2016**, *170*, 110–119.
67. Jara, J.; Holzapfel, E.A.; Billib, M.; Arumi, J.L.; Lagos, O.; Rivera, D. Effect of water application on wine quality and yield in 'Carménère' under the presence of a shallow water table in Central Chile. *Chil. J. Agric. Res.* **2017**, *77*, 171–179.
68. Romero, P.; Fernández-Fernández, J.; Bravo-Cantero, A.; Ayala, M.; Botía, P. Climate Influences on Yield, Berry and Wine Quality in Monastrell Wine Grapes in a Warm Winegrowing Region (Jumilla Area, SE Spain). *Geoinfor. Geostat. Overv.* **2016**, *4*, 3.
69. Conesa, M.R.; Falagán, N.; José, M.; Aguayo, E.; Domingo, R.; Pastor, A.P. Post-veraison deficit irrigation regimes enhance berry coloration and health-promoting bioactive compounds in 'Crimson Seedless' table grapes. *Agric. Water Manag.* **2016**, *163*, 9–18.
70. Viejo, C.G.; Fuentes, S.; Howell, K.; Torrico, D.; Dunshea, F.R. Robotics and computer vision techniques combined with non-invasive consumer biometrics to assess quality traits from beer foamability using machine learning: A potential for artificial intelligence applications. *Food Control* **2018**, *92*, 72–79.
71. Gonzalez Viejo, C.; Fuentes, S.; Torrico, D.; Howell, K.; Dunshea, F.R. Assessment of beer quality based on foamability and chemical composition using computer vision algorithms, near infrared spectroscopy and machine learning algorithms. *J. Sci. Food Agric.* **2018**, *98*, 618–627.
72. Viejo, C.G.; Torrico, D.D.; Dunshea, F.R.; Fuentes, S. Development of Artificial Neural Network Models to Assess Beer Acceptability Based on Sensory Properties Using a Robotic Pourer: A Comparative Model Approach to Achieve an Artificial Intelligence System. *Beverages* **2019**, *5*, 33.

73. Romero, M.; Luo, Y.; Su, B.; Fuentes, S. Vineyard water status estimation using multispectral imagery from an UAV platform and machine learning algorithms for irrigation scheduling management. *Comput. Electron. Agric.* **2018**, *147*, 109–117.
74. Gunaratne, T.M.; Gonzalez Viejo, C.; Gunaratne, N.M.; Torrico, D.D.; Dunshea, F.R.; Fuentes, S. Chocolate Quality Assessment Based on Chemical Fingerprinting Using Near Infra-red and Machine Learning Modeling. *Foods* **2019**, *8*, 426.
75. Fuentes, S.; Chacon, G.; Torrico, D.D.; Zarate, A.; Gonzalez Viejo, C. Spatial Variability of Aroma Profiles of Cocoa Trees Obtained through Computer Vision and Machine Learning Modelling: A Cover Photography and High Spatial Remote Sensing Application. *Sensors* **2019**, *19*, 3054.
76. Fuentes, S.; Tongson, E.J.; De Bei, R.; Gonzalez Viejo, C.; Ristic, R.; Tyerman, S.; Wilkinson, K. Non-Invasive Tools to Detect Smoke Contamination in Grapevine Canopies, Berries and Wine: A Remote Sensing and Machine Learning Modeling Approach. *Sensors* **2019**, *19*, 3335.
77. Gonzalez Viejo, C.; Fuentes, S.; Torrico, D.; Dunshea, F. Non-Contact Heart Rate and Blood Pressure Estimations from Video Analysis and Machine Learning Modelling Applied to Food Sensory Responses: A Case Study for Chocolate. *Sensors* **2018**, *18*, 1802.

Article

Chocolate Quality Assessment Based on Chemical Fingerprinting Using Near Infra-red and Machine Learning Modeling

Thejani M. Gunaratne [1], Claudia Gonzalez Viejo [1], Nadeesha M. Gunaratne [1], Damir D. Torrico [1,2], Frank R. Dunshea [1] and Sigfredo Fuentes [1,*]

1 School of Agriculture and Food, Faculty of Veterinary and Agricultural Sciences, University of Melbourne, Melbourne, VIC 3010, Australia; gunaratnem@student.unimelb.edu.au (T.M.G.); cgonzalez2@unimelb.edu.au (C.G.V.); mgunaratne@student.unimelb.edu.au (N.M.G.); damir.torrico@lincoln.ac.nz (D.D.T.); fdunshea@unimelb.edu.au (F.R.D.)

2 Department of Wine, Food and Molecular Biosciences, Faculty of Agriculture and Life Sciences, Lincoln University, Lincoln 7647, New Zealand

* Correspondence: sfuentes@unimelb.edu.au; Tel.: +61-4245-04434

Received: 1 September 2019; Accepted: 18 September 2019; Published: 20 September 2019

Abstract: Chocolates are the most common confectionery and most popular dessert and snack across the globe. The quality of chocolate plays a major role in sensory evaluation. In this study, a rapid and non-destructive method was developed to predict the quality of chocolate based on physicochemical data, and sensory properties, using the five basic tastes. Data for physicochemical analysis (pH, Brix, viscosity, and color), and sensory properties (basic taste intensities) of chocolate were recorded. These data and results obtained from near-infrared spectroscopy were used to develop two machine learning models to predict the physicochemical parameters (Model 1) and sensory descriptors (Model 2) of chocolate. The results show that the models developed had high accuracy, with $R = 0.99$ for Model 1 and $R = 0.93$ for Model 2. The thus-developed models can be used as an alternative to consumer panels to determine the sensory properties of chocolate more accurately with lower cost using the chemical parameters.

Keywords: sensory; physicochemical measurements; artificial neural networks; near infra-red spectroscopy

---

1. Introduction

Chocolate is a semisolid suspension of fine particles made from sugar, milk powder, milk fat and cocoa in a continuous fat phase. The fruit of *Theobroma cacao* provides the cocoa solids and cocoa butter used for chocolate production. It is a confectionery product that evokes emotional stimuli upon consumption and activates the pleasure centers in the brain [1,2]. The main processing steps of chocolate manufacture are mixing, refining, conching, tempering, molding, and packaging. Particle size reduction takes place in the refining stage to obtain the optimum size of particles, which is important for the texture of chocolate [2]. Viscosity, mouthfeel, and consistency of chocolate are very important properties of chocolate, which are affected by rheological and textural characteristics. The processing conditions and compositions have an influence on the rheological properties of food [3]. To achieve the preferred rheological factors in chocolate with an acceptable texture, several production steps such as mixing, refining, conching, and tempering, are important. These processing parameters affect the viscosity of chocolate [4]. In order to obtain high-quality products, viscosity is considered an important physical parameter in producing chocolate and cocoa products, since it influences the textural properties of chocolate [4]. Moreover, acidity, sweetness, bitterness, color intensity, hardness, and smoothness are the most important parameters which affect the sensory perception of chocolate [5].

Bitterness can be caused by different chemical compounds, such as quinine hydrochloride and 6-n-propyl-2-thiouracil (PROP), and one of its functions is said to be the identification and avoidance of poison [6]. Sodium-ion (Na^+) can be used as a reference to measure saltiness, and it impacts the ion channel of the taste receptor. Acids are responsible for causing sourness, and the hydrogen ion (H^+) activates the taste receptor. Compounds which are responsible for sweetness vary extremely from simple carbohydrates to amino acids and artificial sweeteners [6]. "Umami" is a Japanese term that means "delicious" and is naturally found in food like meats, tomatoes, and mushrooms. It was identified through studies of monosodium glutamate (MSG) and is considered as a flavor enhancer [7].

Research and quality inspection of food products, such as sensory evaluation as well as the determination of physicochemical data is very time-consuming and labor-intensive and may require analytical techniques. Near infra-red (NIR) spectroscopy (NIRS) conforms to 750–2500 nm wavelength, which employs photon energy (hn) within the range of 2.65×10^{-19} to 7.96×10^{-20} J that is a promising technique which may overcome some of the drawbacks of traditional methods [8]. There are several advantages of using NIRS, including being a fast, non-destructive, non-invasive, and universally accepted technique [9]. Several studies have been conducted with NIRS and physicochemical data to quantify various analytes in foods [10]. Recent studies on food products, such as assessment of beer quality using computer vision algorithms, NIRS and machine learning algorithms [11], prediction of pH and total soluble solids in banana using NIRS [12], prediction of canned black bean texture using visible/NIRS [13], and discriminant analysis of pine nuts by NIRS [14] have been conducted. Applications of NIRS to cocoa and chocolate manufacture include quality control of cocoa beans [8], determination of biochemical quality parameters in cocoa [15], rapid determination of sucrose content in chocolate mass [16], and assessment of raw cocoa beans to predict the sensory properties of chocolate [17].

In this study, NIRS was used to gain chemical fingerprinting of chocolate produced using the five basic tastes. Predictive models based on artificial neural networks (ANN) were developed using Matlab® R2018b (Mathworks Inc., Natick, MA, USA). Specific absorbance values of NIR spectra were used as inputs, while physicochemical data (pH, Brix, viscosity, and color) and sensory properties (basic taste intensity) of chocolate were used as targets. The objective of this study was to develop accurate models to predict the quality of chocolate based on chemical, physical, and sensory properties using NIRS and machine learning algorithms.

2. Materials and Methods

2.1 Chocolate Samples

For this study, five different types of chocolate with basic tastes (bitter, salty, sour, sweet, and umami) were used. For the bitter sample, commercially available dark chocolate (70% cocoa) was used. Three concentrations each for the other four basic tastes were produced in the Sensory laboratory at The University of Melbourne, Australia. By tasting at a focus group discussion consisting of sensory professionals at The University of Melbourne, the final concentration of each taste was determined. They commented on the identification and the intensity of each flavor and the concentration which all the participants agreed, was used for final sample preparation. Therefore, salt (4 g), citric acid (0.5 g), sucrose (6 g) and MSG (3.5 g) were added to 100 g of melted compound milk cooking chocolate chips in order to produce salty, sour, sweet and umami samples respectively. These chocolate samples were used for sensory analysis, NIRS, and physicochemical analysis.

2.2. Participants for Sensory Sessions

Panelists (N = 45) were recruited via email invitations from The University of Melbourne, Australia, who volunteered to participate in the sensory assessment of chocolate samples with basic tastes. The panelists had to sign a consent form before participating in the sensory session. The panelists received incentives (chocolate and confectionery products) as an appreciation for participating in the study. The experimental procedure was approved by the Ethics committee of the Faculty of Veterinary and Agricultural Sciences at The University of Melbourne, Australia (Ethics ID 1545786.2).

2.3. Sensory Evaluation

Sensory sessions were conducted in individual booths in the sensory laboratory at The University of Melbourne. Each booth consisted of an integrated camera system controlled by a Bio-sensory application (App) designed for Android tablets (Google; Open Handset Alliance, Mountain View, CA, USA) developed by the sensory group from the School of Agriculture and Food, Faculty of Veterinary and Agricultural Sciences, The University of Melbourne [18]. The intensity of bitterness, saltiness, sourness, sweetness, and umami taste (0—low/7.5—medium/15—high) were assessed using a 15-cm non-structured continuous scale. The temperature of the serving/preparation room was 20 °C, while the temperature of the booths was controlled between 24 and 25 °C. Panelists were served with chocolate samples (two pieces from each taste; each square measuring 1 cm × 1 cm and weighing 7.3 g) on a tray in random order with specific 3-digit random numbers for evaluation. Panelists were asked to cleanse their palate using crackers and water in between all samples.

2.4. pH, Total Soluble Solids (TSS) and Viscosity Measurements

For pH, five readings were taken from five pieces of chocolate (25 readings in total per sample). Initially, 10 g of chocolate was ground and mixed with 100 mL of distilled water and was allowed to settle for 20 min [19]. The pH of the supernatant liquid was measured (25 readings per each taste) using a calibrated bench-top meter (Sper Scientific Ltd., Scottsdale, AZ, USA) at room temperature (23–25 °C). The same method was modified to measure total soluble solids (Brix), and the supernatant liquid obtained from mixing chocolate with distilled water was used to measure the Brix value (25 readings per each taste) using a digital refractometer HANNA HI 96801 (Hanna Instruments Inc., Woonsocket, RI, USA) at room temperature (23–25 °C). In order to determine viscosity, chocolate was melted by microwaving 100 g for 1 min at 800 W [20]. A Brookfield DV1 viscometer with RV7 spindle at 50 RPM (AMETEK Brookfield, Middleboro, MA, USA) was used to measure the viscosity (6 readings per sample) of the melted chocolate at 28–30 °C. About 12–15 min were taken to obtain the measurements.

2.5. Near infra-red Spectroscopy and Color Measurements

A handheld microPHAZIR™ RX Analyzer (Thermo Fisher Scientific, Waltham, MA, USA) was used to measure the NIR spectra of chocolate samples. The instrument was calibrated using a white background before obtaining the measurements as well as after 10–15 samples. This device can be used to measure the absorbance of wavelengths between 1600 and 2396 nm.

In this study, 24 pieces from each type of chocolate were used to measure the spectra. Furthermore, three readings from the top surface and three from the bottom surface of the chocolate piece were obtained, providing 144 readings per sample. The obtained absorbance readings from chocolates manufactured with the five basic tastes were averaged separately in order to plot against wavelength. After considering all the pre-treatments, the second derivative values of absorbance were used for the analysis because they showed the best representation. The Unscrambler X ver. 10.3 (CAMO Software, Oslo, Norway) software was used to plot absorbance values with all wavelengths for the five chocolate samples.

The color parameters using the CIELab color scale were recorded using a StellarNet Inc. EPP2000 (EPP2000-UVN-SR) coupled with an SL1 Filter (StellarNet, Inc., Tampa, FL, USA). In the CIELab color scale, *L* represents lightness and ranges from 0 to 100, *a* depicts the red and green in the positive and negative values, respectively, and *b* represents yellow in the positive values and blue in the negative [21].

2.6. Statistical Analysis and Machine Learning Modeling

An analysis of variance (ANOVA) was conducted to assess significant differences ($\alpha = 0.05$) of the physicochemical data between chocolates with different tastes using Minitab 2017 (Minitab Inc., State College, PA, USA). The mean values and standard deviation of the sensory, chemical, and physical parameters were also calculated and tabulated.

A correlation matrix was also developed to identify the correlations between the sensory (basic taste intensities) and physicochemical properties (pH, Brix, color, and viscosity) of the chocolate samples using a customized code written in Matlab® R2018b (Mathworks Inc. Natick, MA, USA).

As shown in Figure 1, two machine learning models were developed by testing 17 different ANN training algorithms using an automated customized code written in Matlab® R2018b. The 17 algorithms were summed as two backpropagation using Jacobian derivatives, 11 backpropagation using gradient derivatives, and four supervised weight or biased training. An ANN fitting model (Model 1) with three hidden neurons was developed using the normalized NIR readings (−1–1) of chocolate as inputs and normalized values (−1–1) of physicochemical data (pH, Brix, viscosity, and color in CIELab) as targets. The whole NIR spectra (1596–2396 nm) was used to develop the model. The outputs of the Model 1 were used as inputs for the Model 2, while the sensory data (intensities of bitterness, saltiness, sourness, sweetness and umami taste) were used as targets. Both models were developed using a random data division, 70% ($n = 84$) of the samples were used for training, 15% ($n = 18$) for validation with a mean squared error (MSE) performance algorithm and 15% ($n = 18$) for testing using a default derivative function. Data division was randomized similar to Model 1, and 10 hidden neurons were used. From the 17 algorithms (data not shown), the best models corresponded to the Levenberg Marquardt algorithm in both Model 1 and Model 2. The coefficient of determination (R), slope (s), MSE and statistical significance (criteria; $p < 0.05$) were obtained. The p-value was calculated using CoStat ver. 6.45 (CoHort Software, Monterey, CA, USA) software. The performance of the models was evaluated based on the MSE. The most accurate model was selected from the above based on the R and MSE values for each stage (training, validation, testing and overall) and the slope of the overall model. Finally, a normality test (Jarque-Bera Test and DAgostino & Pearson Test) was conducted for the values of the error histogram using a customized code written in Matlab®R2018b to identify whether the errors were normally distributed.

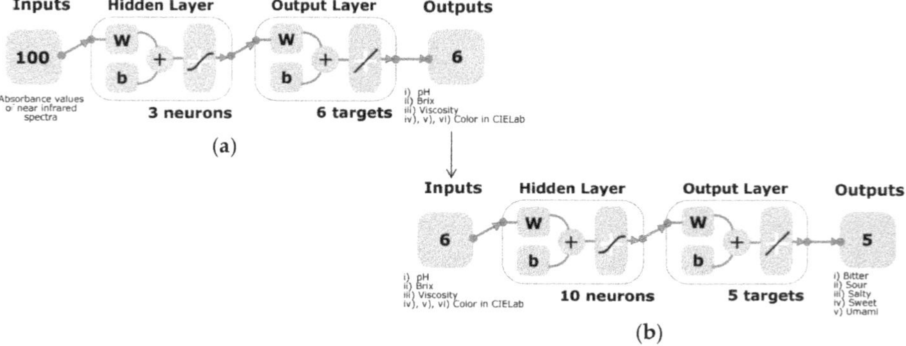

Figure 1. Diagrams with a two-layer feedforward network and tan-sigmoid function in the hidden layer, and a linear transfer function in the output layer for (a) Model 1 constructed with 100 inputs from near-infrared readings, three neurons and six targets related to physicochemical data of chocolate, and (b) Model 2 developed using six inputs obtained from the output of Model 1, ten hidden neurons and five targets related to basic taste intensities of chocolate. For the hidden and output layers, w = weights and b = biases.

3. Results

3.1. Sensory and Chemical Analysis

The mean values and standard deviation of the results obtained from the sensory session (intensities of basic tastes), and physicochemical analysis are shown in Table 1. Significant differences ($p < 0.001$) were found in all the basic taste intensities (bitterness, saltiness, sourness, sweetness, and umami taste) for all five samples. According to the results of ANOVA conducted for the chemical parameters, there was a significant difference ($p < 0.001$) between the different chocolate samples in pH, Brix, and viscosity.

Table 1. Mean and standard deviation values of the sensory, chemical, and color data of the chocolate samples used for this study.

Sample	Bitterness	Saltiness	Sourness	Sweetness	Umami	
Bitter	10.16 ± 3.62 [a]	2.25 ± 2.91 [c,d]	2.35 ± 3.40 [b,c]	4.31 ± 3.34 [c]	3.28 ± 4.13 [c]	
Salty	3.15 ± 3.90 [b]	13.37 ± 2.25 [a]	4.17 ± 4.39 [b]	5.12 ± 3.69 [c]	6.40 ± 4.70 [a,b]	
Sour	2.17 ± 2.85 [b,c]	3.93 ± 3.82 [c]	9.54 ± 4.29 [a]	9.00 ± 3.61 [b]	4.79 ± 3.77 [b,c]	
Sweet	0.99 ± 2.35 [c]	1.84 ± 2.48 [d]	1.15 ± 2.05 [c]	11.95 ± 3.39 [a]	2.56 ± 3.22 [c]	
Umami	2.82 ± 4.07 [b,c]	7.02 ± 3.17 [b]	3.31 ± 3.67 [b,c]	7.85 ± 4.45 [b]	7.43 ± 5.25 [a]	
Sample	**pH**	**Brix**	**Viscosity (cP)**	***L* Value**	***a* Value**	***b* Value**
Bitter	6.40 ± 0.07 [c]	3.90 ± 0.34 [d]	13680 ± 2319 [d]	40.73	12.19	4.37
Salty	6.56 ± 0.07 [b]	6.27 ± 0.89 [a]	23443 ± 618 [a]	65.21	20.47	23.05
Sour	5.42 ± 0.01 [d]	5.64 ± 0.28 [c]	19360 ± 444 [b]	53.57	22.55	23.95
Sweet	6.91 ± 0.40 [a]	6.20 ± 0.19 [a,b]	23600 ± 664 [a]	60.94	23.49	26.02
Umami	6.90 ± 0.05 [a]	5.82 ± 0.58 [b,c]	16747 ± 674 [c]	53.19	30.40	32.30

a–d Means with different letters for each parameter indicate significant differences ($p < 0.05$) by Tukey's studentized Range (HSD) test. ± standard deviation of mean values is stated. Bitterness, saltiness, sourness, sweetness, and umami taste were obtained from a 15-point continuous scale.

3.2. Near Infra-Red Spectroscopy

Figure 2 shows the graphs drawn using the mean values from the replicates of each type of chocolate. Different colors are used to signify each type of chocolate: dark blue for bitter, red for salty, green for sour, light blue for sweet and brown for umami. As seen in this figure, the peak values for all the chocolate samples are within the range of approximately 1700–2350 nm, in which compounds such as water, carbohydrates, proteins, sucrose, lactose, and fat can be found [22].

3.3. Multivariate Data Analysis

The correlation matrix showing the values of the significant correlations ($p < 0.05$) between the chemical and sensory parameters of chocolate is shown in Figure 3. It can be observed that Brix was negatively correlated to bitterness ($R = -0.95$) and positively correlated to the L value ($R = 0.95$) while pH was negatively correlated to sourness ($R = -0.91$). There was also a negative correlation between bitterness and the b value ($R = -0.91$).

Table 2 shows a summary of the statistical data obtained from both Model 1 and 2. Model 1 had a higher correlation coefficient ($R = 0.99$) for the overall stage (Figure 4a). It also had the same performance values for validation and testing stages (MSE = 0.01), which are an indication of no overfitting. The slope values of all the stages were closer to unity ($s \sim 1$). The overall correlation coefficient (R) of Model 2 was 0.93 (Figure 4b). Furthermore, it also had similar performance values for the validation and testing stages (MSE = 0.05) and values closer to the unity ($s \sim 1$) for the slopes of all the stages. Furthermore, data were normally distributed ($p = 1$) in the error histograms of Models 1 and 2 according to the Jarque-Bera Test and the DAgostino & Pearson Test.

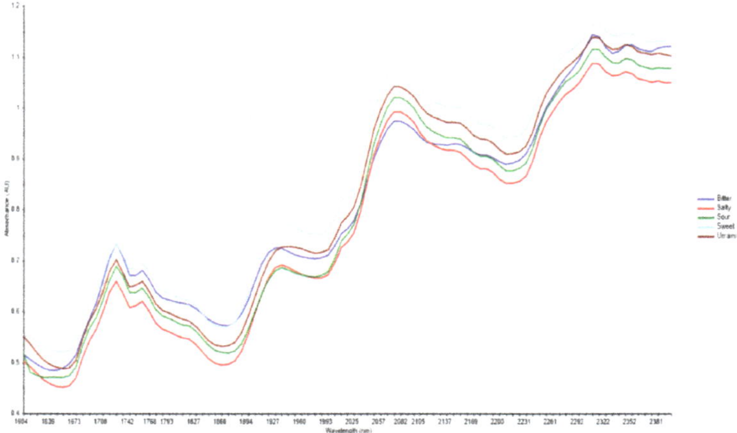

Figure 2. Curves for chocolate samples showing the absorbance (Au) values (y-axis) for specific wavelength (nm) values (x-axis) in the near infra-red spectra for each chocolate sample with basic tastes. A total of 144 absorbance readings per sample were taken, and the curves were drawn by using the mean values.

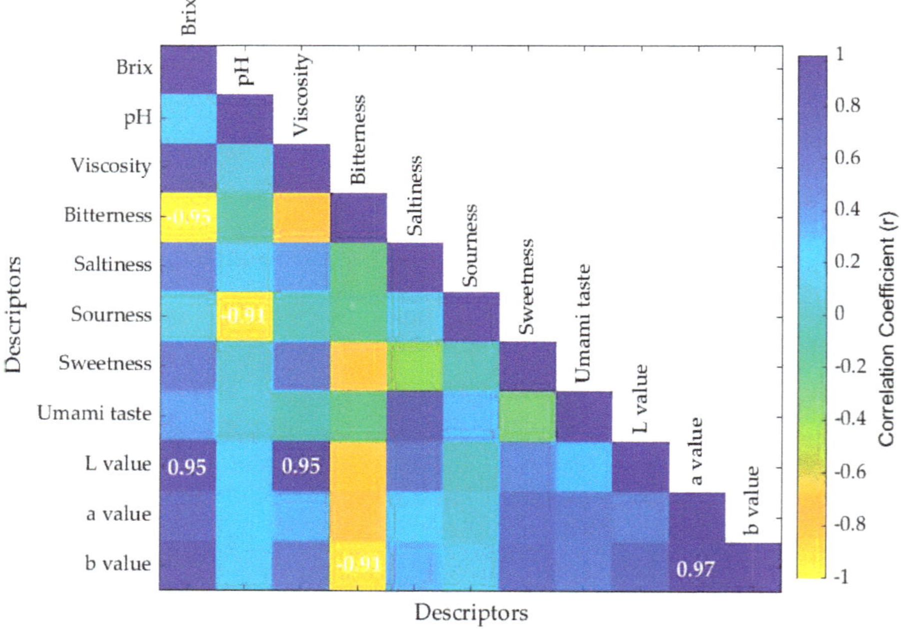

Figure 3. Results from the correlation matrix. Those with the values in the boxes represent the significant correlations ($p < 0.05$). The color bar represents the correlation coefficient (r) with the blue side being positive correlations and the yellow side the negative correlations. The x-axis and y-axis represent the descriptors.

Table 2. Statistical data showing the stage, number of samples, correlation coefficient (R), performance based on mean squared error (MSE), and slope for Model 1 and 2.

Stage	Samples	R	Performance (MSE)	Slope
		Model 1		
Training	84	0.99	0.001	0.98
Validation	18	0.99	0.01	0.99
Testing	18	0.99	0.01	0.97
Overall	120	0.99	0.01	0.98
		Model 2		
Training	84	0.94	0.05	0.88
Validation	18	0.93	0.05	0.91
Testing	18	0.93	0.05	0.86
Overall	120	0.93	0.04	0.88

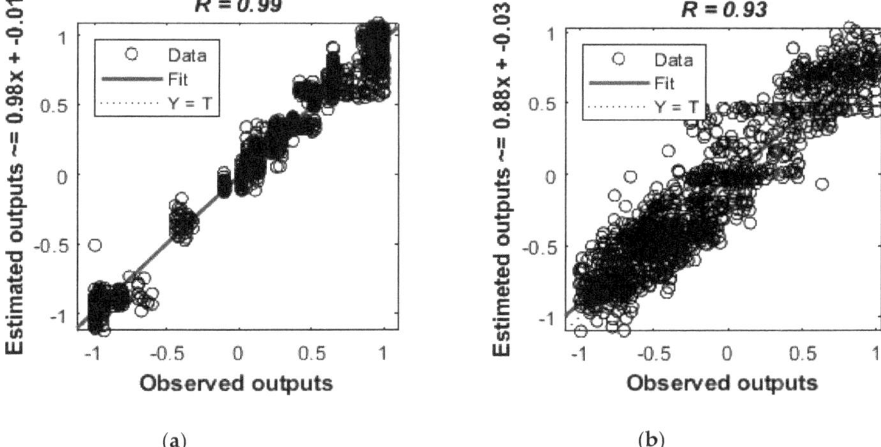

Figure 4. Results of artificial neural networks (ANN); (**a**) Model 1 using physicochemical data as targets and readings of the whole near-infrared wavelength range (1596–2396 nm) as inputs. (**b**) Model 2 using outputs of Model 1 as targets and sensory responses (basic taste intensities) as inputs. The observed values are shown on the x-axis and the estimated values on the y-axis.

4. Discussion

The sour chocolate sample showed the lowest pH value according to the mean values obtained for all samples. This is due to the acidic pH obtained by adding citric acid to the chocolate. Furthermore, bitter chocolate had the lowest Brix value, which is an indicator of the total soluble solids (sugars) of the food. These results are also shown in the correlation matrix (Figure 3), which is discussed later in this paper. Moreover, the bitter chocolate indicated the lowest result for the L value of the color measurements, which shows that it was the darkest sample when compared to others because 0 indicates black and 100 shows white in the L value of the CieLab scale [23].

The NIR curves for all five chocolate samples exhibiting bitter, salty, sour, sweet, and umami taste profiles showed similar trends. The main ingredients of chocolate are cocoa liquor, cocoa butter, sugar, milk powder, and milk fat. These compounds contain peptide and carbohydrate bonds and hence, contribute to the peaks of spectral [24]. According to Figure 2, peaks were observed at 1728 nm and 1761 nm, which match the C-H bond of carbohydrates which is similar to the profile of cocoa butter [24]. The peak around 1940 nm is the water content in the chocolate which may be the water contained in the ingredients used for chocolate production [22]. The NIR curves also showed peaks around 2100 nm and 2310 nm, which show the presence of protein, mainly from the milk powder and cocoa butter in the chocolate. The peak around 2080 nm due to the absorbance of O–H bond in sucrose indicates the presence of sugar in chocolate. Moreover, the peaks around 1759 nm, 2310 nm, and 2343 nm indicate the presence of fats in the chocolate [24]. In food science, qualitative methods play an important role in NIRS analysis based on physicochemical data [25,26]. Hence, these qualitative methods were used in this study to determine the available compounds in the chocolate samples.

Brix is an indicator of the total soluble solids (sugars) in a food product, where the sweetness is high when the value increases [27]. This was consistent with the correlation matrix presented in this study, which showed a negative correlation between Brix and sensory bitterness. The pH is an indicator of acidity in a sample [27]; therefore, sourness decreases with pH as basicity increases, and this was also in accordance with this study, as seen in the correlation matrix as pH had a negative correlation to sourness. Furthermore, bitterness showed a negative correlation to the *b* value, which indicates the reduction of yellow color. This complied with the findings showing that brown color is associated with chocolate products [28]. Hence, the instrumental measurements were correlated to the sensory characteristics. Recently, NIRS applications have been extended by the development of many calibration models along with physicochemical data. In this study, two models with high accuracy were developed using NIRS results, physicochemical data, and sensory intensities to predict the quality of chocolate. These can be used for qualitative or quantitative analysis of food products, and the main methods used for quantitative analysis are principal component regression, step multiple linear regression, partial least squares and ANN [29–34]. In the present study, ANN machine learning models were developed out of the above-mentioned methods.

Chocolate has also been a target for research using NIRS for several purposes. Nutritional parameters have been measured in chocolate using near-infrared diffuse reflectance spectroscopy and neural networks [24]. Similarly to the present study, that study also used ANN models as an alternative to time-consuming chemical methods of analysis. According to the results of the present study, the regression models developed using physicochemical data (pH, Brix, viscosity, color in L*a*b) showed high accuracies with a high correlation coefficient (R^2 = 0.99), meaning that better predictions on chemical factors can be obtained using the developed models.

Both models in this study showed similar MSE values for validation and testing stages, which means that there was no overfitting [35,36]. Due to the high accuracy of Model 2, which can be used to predict the sensory properties of chocolate using chemical and physical parameters (color), it may be used as a fast-screening method to determine the basic taste intensities of chocolate. Using Model 1, we can predict the physicochemical data (pH, Brix, viscosity, and color) of chocolate samples. Furthermore, the sensory properties (intensity of bitterness, saltiness, sourness, sweetness, and umami taste) of chocolate can be predicted using Model 2. Furthermore, this does not require the measurements of NIR but instead, requires the chemical and physical parameters which will be a lower cost compared to near infra-red spectroscopic measurements, considered as a high-cost technique by some researchers [37,38]. Moreover, if an investment can be made on a NIR device, it may be used to obtain all the chemical fingerprinting, physicochemical and the sensory data, which may reduce the time and cost of analysis.

Understanding of food products over regression models for physicochemical data and sensory properties of food can be done using predictive modeling. Research also has been done to assess beer chemometry using novel techniques such as non-linear methods by developing predictive models using PLS and ANN [39]. It showed better predictive models when developed using ANN when compared to PLS. This was also found in the study from Cen and He [29], where they stated that ANN gives better results for some non-linear data than linear approaches. Furthermore, PLS can only generate models for training and validation stages, while ANN can generate for all four stages, which may also be used to find any signs of over or underfitting [11].

NIRS has several benefits when used in food analysis. The spectral measurement takes only 15–90 seconds; hence, NIRS is considered a rapid technique [40]. Furthermore, several samples may be analyzed using one spectral measurement, which is very beneficial in terms of analyzing various indexes. The physical state of the sample also does not matter for NIRS and can be directly tested in the sample chamber. NIRS is also a non-destructive method that is free of chemicals, which is another advantage of this technique [29]. There are few disadvantages in NIRS, such as showing insensitivity to lower concentrations (below 0.1%), the technique being an empirically based quantitative tool and necessity of careful observation for accurate results [41].

5. Conclusions

In this study, the use of NIRS for the prediction of chocolate quality based on chemical, physical, and sensory parameters was reinforced. NIRS and machine learning algorithms can be used to develop accurate prediction models to assess the quality of chocolate. The developed models can be potentially used as a rapid method to obtain physicochemical data and screen the intensity of basic tastes in chocolate based on sensory properties using a minimal amount of laboratory instruments and labor. Further studies may be conducted to improve the quality of models using other physicochemical measurements and sensory properties, which were not considered in this study.

Author Contributions: Conceptualization, T.M.G, C.G.V, D.D.T and S.F; Data curation, C.G.V, D.D.T and S.F; Formal analysis, T.M.G and C.G.V; Funding acquisition, F.R.D; Investigation, T.M.G; Methodology, T.M.G, C.G.V, N.M.G, D.D.T and S.F; Project administration, F.R.D and S.F; Supervision, D.D.T, F.R.D and S.F; Validation, C.G.V, D.D.T and S.F; Writing – original draft, T.M.G; Writing – review & editing, C.G.V, N.M.G, D.D.T, F.R.D and S.F.

Funding: This research was partially funded by the Australian Government through the Australian Research Council (Grant number IH120100053) 'Unlocking the Food Value Chain: Australian industry transformation for ASEAN markets.'

Conflicts of Interest: The authors declare no conflict of interest.

References

1. Cocoa Beans: From Tree to Factory. Available online: https://onlinelibrary.wiley.com/doi/10.1002/9781444301588.ch2 (accessed on 20 September 2019).
2. Afoakwa, E.; Paterson, A.; Fowler, M. Factors influencing rheological and textural qualities in chocolate—A review. *Trends Food Sci. Technol.* **2007**, *18*, 290–298. [CrossRef]
3. Glicerina, V.; Balestra, F.; Dalla Rosa, M.; Romani, S. Rheological, textural and calorimetric modifications of dark chocolate during process. *J. Food Eng.* **2013**, *119*, 173–179. [CrossRef]
4. Servais, C.; Ranc, H.; Roberts, I. Determination of chocolate viscosity. *J. Texture Stud.* **2003**, *34*, 467–497. [CrossRef]
5. Consumer Insight into the Monosodium Glutamate. Available online: https://onlinelibrary.wiley.com/doi/book/10.1002/9781444301588 (accessed on 20 September 2019).
6. Wendin, K.; Allesen-Holm, B.H.; Bredie, W.L. Do facial reactions add new dimensions to measuring sensory responses to basic tastes? *Food Qual. Prefer.* **2011**, *22*, 346–354. [CrossRef]
7. Consumer Insight into the Monosodium Glutamate. Available online: https://scholar.google.com.hk/scholar?hl=zh-CN&as_sdt=0%2C5&as_ylo=2015&q=Consumer+insight+into+the+monosodium+glutamate&btnG= (accessed on 20 September 2019).
8. Hashimoto, J.; Lima, J.; Celeghini, R.M.S.; Nogueira, A.; Efraim, P.; Poppi, R.; Pallone, J.A.L. Quality Control of Commercial Cocoa Beans (*Theobroma cacao* L.) by Near-infrared Spectroscopy. *Food Anal. Methods* **2018**, *11*, 1510–1517. [CrossRef]
9. Pasquini, C. Near Infrared Spectroscopy: Fundamentals, practical aspects and analytical applications. *J. Braz. Chem. Soc.* **2003**, *14*, 198–219. [CrossRef]
10. Wang, S. Infrared Spectroscopy for Food Quality Analysis and Control. *Trends Food Sci. Technol.* **2010**, *21*, 52. [CrossRef]
11. Gonzalez Viejo, C.; Fuentes, S.; Torrico, D.; Howell, K.; Dunshea, F.R. Assessment of beer quality based on foamability and chemical composition using computer vision algorithms, near infrared spectroscopy and machine learning algorithms. *J. Sci. Food Agric.* **2018**, *98*, 618–627. [CrossRef]

12. Ali, M.M. Prediction of total soluble solids and ph in banana using near infrared spectroscopy. *J. Eng. Sci. Technol.* **2018**, *13*, 254.
13. Mendoza, F.; Cichy, K.; Sprague, C.; Goffnett, A.; Lu, R.; Kelly, J. Prediction of canned black bean texture (*Phaseolus vulgaris* L.) from intact dry seeds using visible/near infrared spectroscopy and hyperspectral imaging data. *J. Sci. Food Agric.* **2018**, *98*, 283–290. [CrossRef]
14. Loewe, V.; Navarro-Cerrillo, R.M.; García-Olmo, J.; Riccioli, C.; Sánchez-Cuesta, R. Discriminant analysis of Mediterranean pine nuts (*Pinus pinea* L.) from Chilean plantations by near infrared spectroscopy (NIRS). *Food Control* **2017**, *73*, 634–643. [CrossRef]
15. Krähmer, A.; Engel, A.; Kadow, D.; Ali, N.; Umaharan, P.; Kroh, L.; Schulz, H. Fast and neat—Determination of biochemical quality parameters in cocoa using near infrared spectroscopy. *Food Chem.* **2015**, *181*, 152–159. [CrossRef] [PubMed]
16. Da Costa Filho, P.A. Rapid determination of sucrose in chocolate mass using near infrared spectroscopy. *Anal. Chim. Acta* **2009**, *631*, 206–211. [CrossRef] [PubMed]
17. Davies, A.M.C.; Franklin, J.G.; Grant, A.; Griffiths, N.M.; Shepherd, R.; Fenwick, G.R. Prediction of chocolate quality from near-infrared spectroscopic measurements of the raw cocoa beans. *Vib. Spectrosc.* **1991**, *2*, 161–172. [CrossRef]
18. Fuentes, S.; Gonzalez Viejo, C.; Torrico, D.; Dunshea, F. Development of a Biosensory Computer Application to Assess Physiological and Emotional Responses from Sensory Panelists. *Sensors* **2018**, *18*, 2958. [CrossRef] [PubMed]
19. Homayouni Rad, A.; Roudbaneh, M.; Aref Hosseyni, S. Filled chocolate supplemented with Lactobacillus paracasei. *Int. Res. J. Basic Appl. Sci.* **2014**, *8*, 2026–2031.
20. Vollmer, M.; Möllmann, K.-P.; Karstädt, D. More experiments with microwave ovens. *Phys. Educ.* **2004**, *39*, 346–351. [CrossRef]
21. Conesa, A.; Manera, F.C.; Brotons, J.M.; Fernandez-Zapata, J.C.; Simón, I.; Simón-Grao, S.; Alfosea-Simón, M.; Nicolás, J.M.; Valverde, J.M.; García-Sanchez, F. Changes in the content of chlorophylls and carotenoids in the rind of Fino 49 lemons during maturation and their relationship with parameters from the CIELAB color space. *Sci. Hortic.* **2019**, *243*, 252–260. [CrossRef]
22. Burns, D.A.; Ciurczak, E.W. *Handbook of Near-Infrared Analysis*; CRC Press: Boca Raton, FL, USA, 2007.
23. Erdem, Ö.; Gültekin Özgüven, M.; Berktaş, I.; Erşan, S.; Tuna, H.E.; Karadağ, A.; Özçelik, B.; Güneş, G.; Cutting, S.M. Development of a novel synbiotic dark chocolate enriched with Bacillus indicus HU36, maltodextrin and lemon fiber: Optimization by response surface methodology. *Lebensm. Wiss. Technol.* **2014**, *56*, 187–193. [CrossRef]
24. Moros, J.; Iñón, F.A.; Garrigues, S.; de la Guardia, M. Near-infrared diffuse reflectance spectroscopy and neural networks for measuring nutritional parameters in chocolate samples. *Anal. Chim. Acta* **2007**, *584*, 215–222. [CrossRef]
25. Bucci, R.; Magrí, A.D.; Magrí, A.L.; Marini, D.; Marini, F. Chemical authentication of extra virgin olive oil varieties by supervised chemometric procedures. *J. Agric. Food Chem.* **2002**, *50*, 413–418. [CrossRef] [PubMed]
26. He, Y.; Li, X. Discriminating varieties of waxberry using near infrared spectra. *J. Infrared Millim. Waves Chin. Ed.* **2006**, *25*, 192.
27. Ibarz, A.; Pagán, J.; Panadés, R.; Garza, S. Photochemical destruction of color compounds in fruit juices. *J. Food Eng.* **2005**, *69*, 155–160. [CrossRef]
28. Wei, S.-T.; Ou, L.-C.; Luo, M.R.; Hutchings, J.B. Optimisation of food expectations using product colour and appearance. *Food Qual. Prefer.* **2012**, *23*, 49–62. [CrossRef]
29. Cen, H.; He, Y. Theory and application of near infrared reflectance spectroscopy in determination of food quality. *Trends Food Sci. Technol.* **2007**, *18*, 72–83. [CrossRef]
30. Gomez, A.H.; He, Y.; Pereira, A.G. Non-destructive measurement of acidity, soluble solids and firmness of Satsuma mandarin using Vis/NIR-spectroscopy techniques. *J. Food Eng.* **2006**, *77*, 313–319. [CrossRef]
31. Ni, Y.; Zhang, G.; Kokot, S. Simultaneous spectrophotometric determination of maltol, ethyl maltol, vanillin and ethyl vanillin in foods by multivariate calibration and artificial neural networks. *Food Chem.* **2005**, *89*, 465–473. [CrossRef]
32. Geesink, G.; Schreutelkamp, F.; Frankhuizen, R.; Vedder, H.; Faber, N.; Kranen, R.; Gerritzen, M. Prediction of pork quality attributes from near infrared reflectance spectra. *Meat Sci.* **2003**, *65*, 661–668. [CrossRef]

33. Aske, N.; Kallevik, H.; Sjöblom, J. Determination of saturate, aromatic, resin, and asphaltenic (SARA) components in crude oils by means of infrared and near-infrared spectroscopy. *Energy Fuels* **2001**, *15*, 1304–1312. [CrossRef]
34. Blanco, M.; Coello, J.; Iturriaga, H.; Maspoch, S.; Pages, J. Calibration in non-linear near infrared reflectance spectroscopy: A comparison of several methods. *Anal. Chim. Acta* **1999**, *384*, 207–214. [CrossRef]
35. Goodfellow, I.; Bengio, Y.; Courville, A.; Bengio, Y. *Deep Learning*; MIT Press: Cambridge, UK, 2016; Volume 1.
36. Beale, M.H.; Hagan, M.T.; Demuth, H.B. *Neural Network Toolbox™ User's Guide*; The MathWorks, Inc.: Natick, MA, USA. Available online: www.mathworks.com (accessed on 20 September 2019).
37. Yan, L.; Bai, Y.; Yang, B.; Chen, N.; Tan, Z.A.; Hayat, T.; Alsaedi, A. Extending absorption of near-infrared wavelength range for high efficiency CIGS solar cell via adjusting energy band. *Curr. Appl. Phys.* **2018**, *18*, 484–490. [CrossRef]
38. Quelal Vásconez, M.; Pérez Esteve, É.; Arnau Bonachera, A.; Barat, J.; Talens, P. Rapid fraud detection of cocoa powder with carob flour using near infrared spectroscopy. *Food Control* **2018**, *92*, 183–189. [CrossRef]
39. Cajka, T.; Riddellova, K.; Tomaniova, M.; Hajslova, J. Recognition of beer brand based on multivariate analysis of volatile fingerprint. *J. Chromatogr. A* **2010**, *1217*, 4195–4203. [CrossRef]
40. Wan, L.; Sun, H.; Ni, Z.; Yan, G. Rapid determination of oil quantity in intact rapeseeds using near-infrared spectroscopy. *J. Food Process Eng.* **2018**, *41*, e12594. [CrossRef]
41. McClure, W.F. Near-Infrared Spectroscopy The Giant is Running Strong. *Anal. Chem.* **1994**, *66*, 42A–53A. [CrossRef]

Article

Comparative Study of Several Machine Learning Algorithms for Classification of Unifloral Honeys

Fernando Mateo [1,*], Andrea Tarazona [2] and Eva María Mateo [3]

[1] Department of Electronic Engineering, ETSE, University of Valencia, 46100 Burjasot, Spain
[2] Department of Microbiology and Ecology, University of Valencia, 46100 Burjasot, Spain; Andrea.tarazona@uv.es
[3] Department of Microbiology, School of Medicine, University of Valencia, 46010 Valencia, Spain; Eva.mateo@uv.es
* Correspondence: Fernando.mateo@uv.es

Abstract: Unifloral honeys are highly demanded by honey consumers, especially in Europe. To ensure that a honey belongs to a very appreciated botanical class, the classical methodology is palynological analysis to identify and count pollen grains. Highly trained personnel are needed to perform this task, which complicates the characterization of honey botanical origins. Organoleptic assessment of honey by expert personnel helps to confirm such classification. In this study, the ability of different machine learning (ML) algorithms to correctly classify seven types of Spanish honeys of single botanical origins (rosemary, citrus, lavender, sunflower, eucalyptus, heather and forest honeydew) was investigated comparatively. The botanical origin of the samples was ascertained by pollen analysis complemented with organoleptic assessment. Physicochemical parameters such as electrical conductivity, pH, water content, carbohydrates and color of unifloral honeys were used to build the dataset. The following ML algorithms were tested: penalized discriminant analysis (PDA), shrinkage discriminant analysis (SDA), high-dimensional discriminant analysis (HDDA), nearest shrunken centroids (PAM), partial least squares (PLS), C5.0 tree, extremely randomized trees (ET), weighted k-nearest neighbors (KKNN), artificial neural networks (ANN), random forest (RF), support vector machine (SVM) with linear and radial kernels and extreme gradient boosting trees (XGBoost). The ML models were optimized by repeated 10-fold cross-validation primarily on the basis of log loss or accuracy metrics, and their performance was compared on a test set in order to select the best predicting model. Built models using PDA produced the best results in terms of overall accuracy on the test set. ANN, ET, RF and XGBoost models also provided good results, while SVM proved to be the worst.

Keywords: machine learning; unifloral honeys; botanical origin; physicochemical parameters; classification

Citation: Mateo, F.; Tarazona, A.; Mateo, E.M. Comparative Study of Several Machine Learning Algorithms for Classification of Unifloral Honeys. *Foods* **2021**, *10*, 1543. https://doi.org/10.3390/foods10071543

Academic Editor: Sigfredo Fuentes

Received: 31 May 2021
Accepted: 29 June 2021
Published: 3 July 2021

1. Introduction

Honey is a natural food appreciated worldwide with high nutritional value that provides many health benefits [1,2]. Honey is defined by the European Union (EU) as "the natural sweet substance produced by *Apis mellifera* bees from the nectar of plants or from secretions of living parts of plants or excretions of plant-sucking insects or the living parts of plants, which the bees collect, transform by combining with specific substances of their own, deposit, dehydrate, store, and leave in honey combs to ripen and mature" [3]. The EU regulations concerning honey are included mainly in the 2001/110/EC Council Directive [4], further amended by the 2014/63/EU Directive [5]. The composition criteria for honey placed on the market or used in any product intended for human consumption are stated in Annex II of [4]. The Codex standard for honey adopted by the Codex Alimentarius Commission in 1981 was revised in 1987 and 2001, has served as a basis for national legislations in some countries and has voluntary application [6]. Honey is a very rich food product that contains water, sugars (mainly fructose and glucose, but also di- and trisaccharides), hydroxymethyl furfural and other compounds at low levels such as minerals, amino acids, proteins (including enzymes), aromatic acids, esters, aroma components and flavonoids [1,2]. Naturally, bees forage the flowers they can access. Hence, the honey produced mostly has a blend of flavors and is commonly sold in the market simply as honey or mixed-flower honey. However, when the nectar is taken predominantly from a single type of flower, the honey produced has characteristic organoleptic properties, adding to its commercial value. Many consumers appreciate these particular sensorial properties very much, which increase these honeys' price with respect to other types of honey. Moreover, honeydew honeys are especially appreciated by consumers in Central Europe (Germany, Switzerland and Austria). Denominations of botanical origin are extensively used on the honey market as they offer consumers the choice among a variety of different typical products, paying prices depending on local consumer preferences [7]. The existing international norms and regulations do not specify the characteristics of unifloral honeys, although limits for moisture content, sugar content or electrical conductivity are different for honeys originated from some botanical origins [2]. Authentication of food products is of great concern in the context of food safety and quality. In recent years, interest in honey authenticity in relation to botanical or geographical origin and adulteration has increased. Due to the huge variety of different floral sources normally attainable by bees for foraging and to the great diversity within plant species, which is influenced by the climatic and growing conditions, the parameters used for characterizing unifloral honeys do not exhibit typical values but are defined in rather large, often overlapping ranges [7]. The differences observed in honey composition depend on a variety of factors, such as the region, season, nectar source, beekeeping practices and harvest period [8].

Classically, the determination of the botanical origin of honeys has been performed by melissopalynological methods. The fundamentals of this methodology were established many years ago [9,10], but it has been used for years. Usually, honey is considered mainly from one plant if the pollen frequency of that plant is >45%. Pollen grains from anemophilous plants and plants with nectarless flowers are excluded in the calculation of the percentages. Moreover, pollen grains from some species are under- or over-represented in relation to the nectar their flowers yield. For unifloral honeys with under-represented pollen, the minimum percentage of the taxon that gives the honey its name ranges 10–30%; for those with over-represented pollen, the minimum percentage can be 80–90%. This technique is useless in the case of honey filtration. Notwithstanding, interpretation of pollen analysis data may be difficult in some cases, and the counting and identification of pollen grains depend greatly on the skill and performance of the analyst [11]. Sensory properties (color, aroma, flavor) can help to ascribe a honey sample to a given botanical origin, but, due to subjectivity, well-trained personnel are needed. However, the sensory properties of a honey can vary with time and thermal treatment while maintaining the floral origin. Organoleptic properties have been considered, together with pollen analysis, key to performing the classification of unifloral honeys. Methods based on physicochemical properties of honey have been developed for the accurate classification of these honeys with the help of suitable statistical treatments [11]. Generally, no single parameter has proved useful to characterize the botanical origin of honey, except the methyl anthranilate content, which is characteristic of citrus honey [12,13]. Assayed parameters have been honey color (measured using CIE-1931 xyL or CIE-1976 LAB chromatic coordinates) [14], the carbohydrate profile [15–17], volatile organic compounds [16,18–22] or the amino acid profile [18,23].

Even when differences among honeys from distinct botanical sources are found using only a profile of a single class of compounds (sugars, amino acids, volatiles, etc.) or characteristics, a thorough characterization of the botanical origin of honeys is not achieved. Thus, sets of different parameters, either physicochemical or sensorial, or both, sometimes with the pollen spectrum and usually involving statistical (chemometric) techniques, such as cluster analysis, principal component analysis (PCA) and linear discriminant analysis (LDA), have been considered. Parameters tested together with this aim have been water content, pH, acidity, electrical conductivity, some carbohydrates, color, volatile compounds, amino acids, phenolic compounds, mineral elements, etc. [11,22–26]. Even when the chemical composition of honey is associated with its botanical and geographical origin, some processes, such as heating, storing or the extraction techniques, can alter the initial volatile composition [22], which affects the volatile fingerprint of unifloral honeys and hence organoleptic properties. Other classification approaches lie in the use of nondestructive techniques applied to honey samples. In this way, attenuated total reflectance Fourier-transform infrared spectroscopy (ATR-FTIR) of unifloral honeys is a technique that, after treatment by PCA and further treatment of the principal components by means of a machine learning (ML) algorithm such as support vector machine (SVM), proved useful for the characterization of honey origin [27]. The potential application of other spectroscopy techniques such as visible–near-infrared (VIS–NIR) hyperspectral imaging for the detection of honey flower origin using ML techniques has been reported [28]. PCA was used for dimensionality reduction before ML treatment using three ML algorithms, namely, radial basis function (RBF) network, SVM and random forest (RF), to predict honey floral origin. Furthermore, FT-Raman spectroscopy has shown to be a simple, rapid and nondestructive technique that, in combination with proper PCA or LDA models, could be successfully adopted to identify the botanical origins of some honey types [29–31]. The same technique resulted in being useful to detect adulterations of pure beeswax with paraffin or microcrystalline waxes [32]. Nuclear magnetic resonance (NMR) was used for the estimation of the botanical origin of honeys. Due to the complex nature of NMR data, multivariate analysis has been applied to extract the useful information [33]. The application of an electronic nose (E-nose) to parametrize the odor compounds in the form of numeric resistance and further treatment by k-nearest neighbor (k-NN) has been reported [34]. A commercial electronic tongue including seven potentiometric sensors has been applied for the classification of honeys. Botanical classification was performed by PCA, canonical correlation analysis (CCA) and artificial neural network (ANN) modeling on samples of acacia, chestnut and honeydew honeys [35]. The ML algorithms applied to the authentication of the botanical origin of honeys are ANN [35,36], classification and regression trees (CART) [37], k-NN, SVM or RF [27,28,34]. However, the usage of ML techniques is not popular in honey research, and mixed approaches including classical statistics together with ML have been applied in some studies, as indicated in a recent review [38].

The aim of the present study was to carry out a comparative analysis of the application of some ML algorithms to find the most useful to accurately classify rosemary, citrus, lavender, sunflower, heather, eucalyptus and forest unifloral honeys harvested in Spain on the basis of some physicochemical properties (pH, moisture, electrical conductivity, sugars) and color. The classifier algorithms used for this goal were penalized discriminant analysis (PDA), high-dimensional discriminant analysis (HDDA), shrinkage discriminant analysis (SDA), nearest shrunken centroids (PAM), partial least squares (PLS) or decision trees (5.0 tree), extremely randomized trees or Extra Trees (ET), k-NN, SVM, RF and extreme gradient boosted tree (XGBoost).

2. Materials and Methods

2.1. Honey Samples

The analyzed Spanish honey samples were obtained from beekeepers and traders before processing. They belonged to seven unifloral origins, most of which are very appreciated by consumers worldwide or in countries of Central Europe. They were rosemary (*Rosmarinus officinalis* L.), orange blossom (*Citrus* spp.), lavender (*Lavandula latifolia, L. angustifolia, L. vera*), sunflower (*Helianthus annuus* L.), heather/bell heather (Ericaceae, mainly *Erica* spp. and *Calluna vulgaris*), eucalyptus (*Eucalyptus globulus* and *E. camaldulensis*) and forest honeys. Honeys were harvested in different regions of Spain, excluding the Balearic and Canary Islands. The approximated coordinates of the areas related to honey harvest are longitude 3°19′ E–7° W and latitude 36° N–43° N. Orange blossom honey was harvested mainly in eastern Spain (Valencian Community) and some provinces of Andalucia. Rosemary and lavender honeys were harvested mainly in central and southeastern Spain (Castilla-la Mancha, Castilla-Leon) during spring (rosemary) and summer (lavender). Heather and bell heather (in the following, heather) honey was harvested in many Spanish regions during March or September. Forest honey was mainly honeydew honey from holm oak (*Quercus ilex* L., *Q. rotundifolia* Lam., *Q. bellota* Desf.) grown in western Spain (Extremadura and Salamanca), and it was harvested during August/September or later. Eucalyptus honey was harvested mainly in provinces located in western Spain (Huelva, Extremadura) and northwestern Spain during September–October. Sunflower honey was collected during summer in southern/central Spain (Andalucía and Castilla-La Mancha). Samples were collected in different years from 2010 to 2014.

Samples were screened by microscopic and sensory analysis (color, aroma, taste) assessment as soon as they arrived at the laboratory. When analysis had to be delayed for more than four weeks, they were stored at −20 °C; otherwise, they were stored at 4–6 °C in the dark.

The samples were assessed microscopically for pollen and honeydew elements (HDE), which are mainly unicellular algae, fungal spores and hyphae. HDE/pollen (from nectariferous plants) ratios higher than three are required for honeydew honeys according to Louveaux et al. [10]. However, an HDE/pollen ratio of 1.5 ± 1.2 (0.3–4) was reported in 167 honeydew honeys from different places in Europe [13]. Studies performed in Spain have found rather low values for such index in oak honeydew honey [39]. Melissopalynology seems not to be useful for classification of Spanish oak honeydew honeys [40]. Other required parameters associated with honeydew honeys are electrical conductivity values >800 µs/cm [4,6] and pH values >4.3 [13], besides acceptable sensory assessment (dark amber color, characteristic taste, lack of crystallization tendency). It is known that in rosemary, lavender and citrus honeys, pollen from the flowers of these plants is not dominant. After screening, some samples were rejected as unifloral. The number of honey samples collected before the initial screening and the number of samples eventually selected for all physicochemical analyses and statistical treatments are indicated in the following relation, where the selected samples are between parenthesis: 27(13) from rosemary, 31(13) from heather, 35(16) from orange blossom, 33(16) from forest, 19(14) from lavender, 23(14) from eucalyptus and 33(14) from sunflower.

2.2. Microscopical Analysis

Microscopical analysis of honey sediment was achieved according to the methods of melissopalinology [10] and the Spanish official methods of analysis for honey [41]. Slides were prepared without acetolysis. Briefly, graduated conical centrifuge tubes containing 10× *g* of homogenized honey solved in 20 mL of dilute sulfuric acid were centrifuged for 10 min at 2500 rpm. The supernatant was discarded, and the sediment was washed twice with 10 mL of distilled water and centrifuged. After discarding the supernatant, the sediment was homogenized, and an aliquot was placed on a glass slide, sprouted over an area of 4 cm^2, dried at 40 °C and mounted with stained glycerin-gelatin. Pollen grains were identified by light microscopy with the aid of non-acetolyzed pollen collection and microphotographs from specialized studies. Usually, 350–500 grains were counted, and they were classified in the following frequency classes: dominant pollen (>45% of the pollen grains counted); secondary pollen (16–45%); important minor pollen (3–15%); minor pollen (1–3%); and present (<1%). For forest honey, HDE were counted apart from pollen grains.

2.3. Electrical Conductivity

Electrical conductivity was measured at 20.0 ± 0.1 °C in a 20% (w/v) solution of honey (dry matter basis) in deionized water with electrical conductivity of <1 µs/cm [41] using a Crison model 525 conductimeter (Crison Instruments, Barcelona, Spain). The cell was previously calibrated at 20.0 °C with a 0.01 M KCl solution. Measurements were carried out in quintuplicate.

2.4. Water Content

Water content was determined at 20.0 °C by refractometry according to [41]; this method matches the AOAC 969.38B method [42] cited in [6]. A Bellingham and Stanley standard model Abbe-type refractometer previously calibrated and connected to a thermostatic bath was used. The Chataway tables revised by Wedmore were used to convert refraction indices to percentage of water [41]. Measurements were carried out in triplicate.

2.5. pH Measurement

Measurements of pH were performed at 20.0 ± 0.1 °C in a 10% (*w*/*v*) solution of honey in freshly boiled distilled water using a Crison micropH 2000 pH-meter (Crison Instruments, Barcelona, Spain). The pH-meter was calibrated with buffers of pH 6.50 and 3.00 just before measurements, which were conducted in triplicate.

2.6. Color

Honeys were liquefied if needed by heating at 50–60 °C, in the case of crystallized samples, and then left to cool at room temperature. Color of liquid honeys was determined by measurement of transmittances at 30 selected wavelengths, on a Shimadzu UV-vis 240 spectrophotometer fitted (Shimadzu Co., Tokyo, Japan). The x, y and L chromatic coordinates from the CIE-1931 (xyL) color system [43] were calculated from the tristimulus values [14]. Transmittance measurements were conducted in triplicate.

2.7. Sugars

Sugars were determined by gas chromatographic (GC) separation of the trimethylsilyl (TMS) derivatives (TMS oximes and TMS ethers in the case of non-reducing sugars) in an OV-17 packed column on a Perkin-Elmer Sigma 3 gas chromatograph equipped with a flame ionization (FID) detector (Perkin-Elmer Co., Norwalk, CT, USA) [15,41]. The sugars determined were fructose, glucose, sucrose, kojibiose, isomaltose and maltose. The last disaccharide includes not only maltose but also nigerose and turanose due to peak overlapping. The fructose/glucose and glucose/water ratios were also calculated. The trisaccharides raffinose, erlose and melezitose were estimated, but due to uncertainty in determination or lack of detection, they were not used in classification algorithms. Analyses were conducted in triplicate.

2.8. Classification Using Statistical Multivariate and Machine Learning Algorithms

Once the dataset with the values for all the inputs was obtained, several algorithms were applied to compare their performance to achieve the classification of the samples as accurately as possible. First, in an exploratory analysis, a correlation matrix was obtained. Then, PCA, a well-known unsupervised statistical method, was used to examine the data; it identifies orthogonal directions of maximum variance in the dataset, in decreasing order, and projects the data in a lower-dimensionality space formed of a subset of the components with the highest variance. The orthogonal directions are principal components, which are linear combinations of the original input variables. It is a method for feature reduction and transforms the original independent variables into new axes. PCA helps to identify patterns in data and express them in such a manner to indicate their similarities and differences [27,38]. The k-means algorithm was also applied to the dataset to find clusters among the samples.

The first supervised classification approach was to consider most of the dataset (70%) randomly selected for a training task using 10-fold cross-validation to validate the models and, after finishing training and optimizing the parameters, to utilize the remaining dataset (30%) to test the ability of the best model to accurately classify the samples into their a priori labeled parent classes. Statistical multivariate algorithms applied to this goal were the following: penalized discriminant analysis (PDA) [44], which is a penalized version of Fisher linear discriminant analysis (LDA); the K-NN algorithm, which assumes the similarity between the new sample and available samples within a K distance and places a new sample into the class that is most similar among the available classes. It is non-parametric, i.e., it does not make any assumption on the underlying data. In fact, the used algorithm was a weighted version of K-NN included in the KKNN method of "caret" [45,46]. Other tested classifiers have been high-dimensional discriminant analysis (HDDA) [47], which is a model-based discriminant analysis method that assumes that each class of the dataset resides in a proper Gaussian subspace that is much smaller than the original one, and the function calculates the parameters of each subspace to predict the class of new observation of this kind; nearest shrunken centroids (NSC) [48,49], also known as prediction for microarrays (PAM); C5.0 tree, which is a popular implementation of decision trees; partial least squares (PLS), a multivariate linear regression method that can deal with a large number of predictors, a small sample size and high collinearity among predictors and acts by forming linear combinations of the predictors in a supervised manner; extremely randomized trees (ET) [50]; and shrinkage discriminant analysis (SDA) [51], an algorithm that determines a ranking of predictors by computing CAT scores (correlation-adjusted t-scores) between the group centroids and the pooled mean. Variables were preprocessed and scaled before treatment. Several metrics were used for tuning the algorithm parameters during training/validation. They were logistic loss (log loss), also known as cross-entropy loss, which is a classification loss function that tends to a minimum (the lower limit is zero but there is not a high limit) as model performance increases, meaning that the objective is to minimize the expected loss or risk; accuracy, which increases (the upper limit is 1) as the performance of the classifier increases, that is, when labeled samples are included into their a priori known classes; area under the curve (AUC), which computes the area under the receiver operator characteristic (ROC) curve and also increases (value range 0-1) with classifier performance; Cohen's kappa; sensitivity; specificity; and precision, among others. However, log loss was primarily used to measure the performance to build the best model across training/cross-validation, except when this function is not implemented.

Other ML algorithms included in the comparison were ANN (neuralnet library) [52], SVM with linear kernels (SVM library), which is a well-known classifier [53,54], RF (randomForest library) [55], XGBoost (XGBTree library) [56] and extremely randomized trees (ET) [50]. The software used was R and the "classification and regression training" (caret) package [46]. The confusion matrices express the number of samples accurately classified into their parent class or otherwise. To conduct a fair comparison of the different algorithms, the dataset partitions into training and test sets were identical.

3. Results

3.1. Honey Dataset

For the selected honey samples after microscopy analysis and sensory assessment, a dataset was built using the mean values of each determination. The dataset is summarized in Figures S1 and S2. After microscopic analysis, pollen count was related to nectariferous plants. The box plots of percentages of pollen from the taxa that give the names to the studied unifloral honeys are shown in Figure S2g. Selected samples were as follows: Rosemary honeys that had 20–77% pollen from *R. officinalis* were considered acceptable as it is known as an under-represented pollen. Orange blossom or citrus honeys had a percentage of *Citrus* spp. pollen in the range 10–46%, except in one sample (80%). Citrus honeys are considered unifloral if the pollen of *Citrus* spp. is >10% because it is considered as under-represented. Lavender honeys sowed a percentage of *Lavandula latifolia* or *L. spica* in the range 15–68%. Pollen from *L. stoechas* was usually absent. Additionally, in this honey class, the pollen is considered under-represented. Sunflower honeys had pollen of *H. annuus* in the range 31-82%. Eucalyptus honeys contained 82-98% pollen of *Eucalyptus* spp. High counts in this case are usual because *Eucalyptus* pollen is over-represented. Heather honeys encompassed pollen from *Erica* spp. in the range 48–80% (Figure S2g). For forest honey, which is mainly honeydew honey, pollen counts of *Quercus* spp., although always present, are of no interest, as previously commented, because their flowers are non-nectariferous, but they were always examined microscopically for HDE and the presence of pollen from other taxa. HDE presence was scarce. Concerning organoleptic properties, rosemary and orange honeys displayed a light amber color and had a characteristic aroma and taste. Lavender and eucalyptus honeys were light amber but darker than orange or rosemary honeys and had a characteristic aroma and taste. Sunflower honeys had a yellow characteristic and a bright golden-amber color, with a yellow hue and slight tart aroma, and crystalized easily, producing fine crystals. Heather honeys were amber/dark amber with a reddish hue and had a characteristic intense aroma and sour taste and a tendency to crystallize. Forest honeys were also dark amber/dark, had an intense flavor, were slightly bitter and sour and remained liquid even in cool conditions for months.

Figures S1 and S2 show the large variability of the data. Some rosemary and citrus honeys had a high moisture percentage, and the lower water levels were found in eucalyptus and forest honeys, while the remaining honey types exhibited intermediate water contents (Figure S1a). All lavender honeys were well below the 15% limit for sucrose established in the EU Council directive [4]. Forest and heather honeys showed the highest values of electrical conductivity, followed by eucalyptus and lavender/sunflower honeys, while rosemary and citrus displayed the lowest values for this parameter (Figure S1). The pH was also higher in forest and heather honeys than in the remaining honeys. The highest contents of fructose and glucose and the highest glucose/water ratio were found in sunflower honeys, which also had the lowest fructose/glucose ratio; forest honeys showed the lowest levels of both fructose and glucose. On the contrary, maltose, isomaltose and kojibiose contents and the fructose/glucose ratio reached the highest values in forest honeys (Figures S1 and S2). Concerning the color parameters, the largest x values were observed in heather honeys followed by forest honeys, and the minimum x values were observed in rosemary and citrus honeys. However, heather honeys had the lowest mean value for the y and L chromatic coordinates. The largest mean y value was exhibited by sunflowers honeys, and the largest mean L value was observed in citrus honeys (Figure S2).

3.2. Statistical and ML Algorithms

A multivariate statistical study of the dataset was carried out initially. Variables were centered and scaled before statistical treatments. Correlations, PCA and data clustering were performed. A diagram including correlations between all the variables is shown in Figure 1. A score plot of the two principal components can be observed in Figure 2. PC1 and PC2 account for 45.27% and 20.46% of the variance, respectively (overall 66.73%). Heather and forest honeys spread along the positive side of PC1. Sunflower honeys extend on the negative side of PC1, but on the positive side of PC2. All citrus and most rosemary honeys are on the negative side of both PC1 and PC2. Eucalyptus honey samples spread on the positive side of PC2, and most of them are on the negative side of PC1, while most lavender honeys fall on the negative side of PC1, but they spread on both the positive and negative sides of PC2.

Another unsupervised way to explore the dataset, k-means clustering [57,58], was run to partition the data into a number of clusters using the library "factoextra" in R. All the input variables were taken into account. Two clusters of sizes 70 and 30 were obtained on the basis of the maximum average silhouette width (Figure 3). However, this number of clusters is an estimate and does not mean that only two clusters may exist. The mean values for the variables in each of these two clusters (corresponding to the centroids) are listed in Table S1. Cluster 2 is smaller in size than cluster 1 and is higher than cluster 1 in mean values of electrical conductivity, pH, disaccharides (except sucrose), fructose/glucose ratio and the x chromatic coordinate. When comparing Figures 2 and 3b, cluster 2 seems to encompass forest and heather honeys and cluster 1 the remaining honeys. Forcing the k-means clustering to display seven groups on a two-dimensional plot leads to highly overlapped clusters (Figure S3). The relative importance of the variables was tested using an RF model as a reference (Figure S4). The most important variables are electrical conductivity, the chromatic coordinates, water content, fructose and glucose. The less important variables are glucose/water and fructose/glucose ratios. The Boruta package [59] was applied, and it considered that no variables had to be removed regardless of their relative importance.

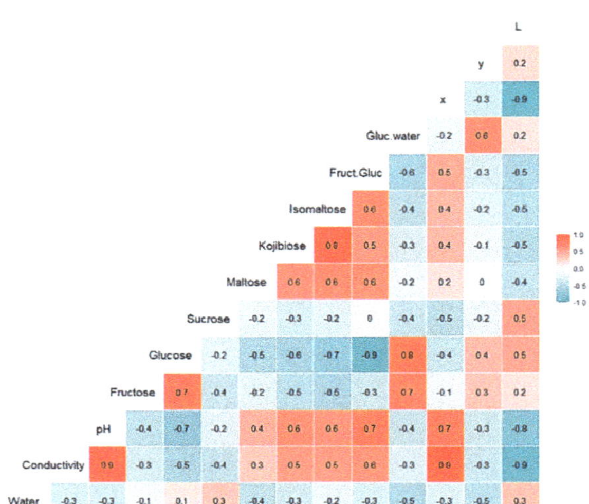

Figure 1. Correlation chart among the 14 predictor variables for the whole dataset. The number in each square is rounded to one figure. The color scale at the right indicates color meaning. Red color means positive correlation; blue color means negative correlation.

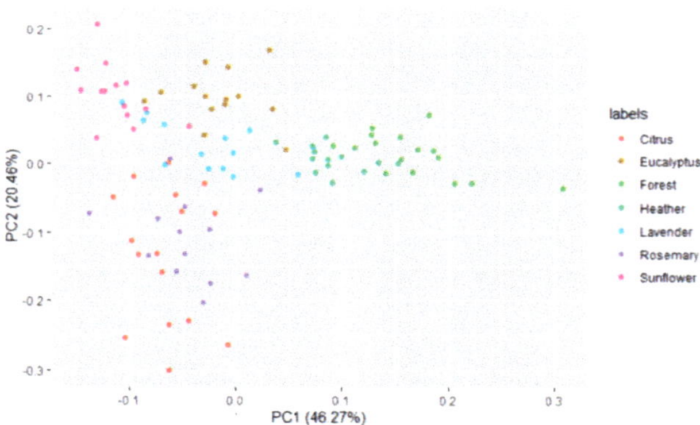

Figure 2. Principal component score plot based on the 14 variables of 100 honey samples according to the botanical origins.

Using the approach of supervised modeling, different classifier algorithms were applied to the dataset, which was divided into a training set (70%) and a test set (30%). Ten-fold cross-validation was applied during training with four repetitions. Various metrics (log loss, accuracy, AUC, kappa, sensitivity, specificity, precision, etc.) can be used during training to tune the key parameters of the algorithms in order to find the best ones. The absolute values of metrics vary when training is repeated. Among them, the log loss metric was usually chosen to select the optimal model using the smallest value. For KKNN or KNN (as weights were not relevant), the final value of the tuning parameters used for the optimized model was kmax (maximum number of neighbors) = 5 (Table 1).

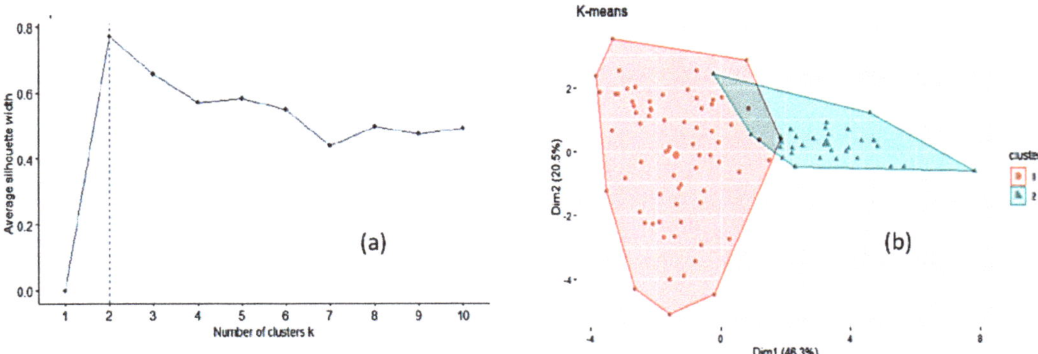

Figure 3. Optimal number of clusters by k-means using the average silhouette width (**a**) and clustering of honey samples by k-means algorithm in two clusters where the two largest symbols are the centroids of each cluster (**b**).

Table 1. Model optimization using the classifier algorithms. Log loss values are means of 10-fold cross-validation.

Algorithm	Tuning Parameter	Mean Log Loss Values
KKNN	Kmax = 5	0.8339319
	Kmax = 7	0.9017721
	Kmax = 9	0.9808674
PDA	Lambda = 1	0.5689435
	Lambda = 0.0001	0.5687306
	Lambda = 0.1	0.4611719
HDDA	Thershold = 0.05	0.4360396
	Thershold = 0.175	1.3732500
	Thershold = 0.300	1.0080708
SDA	Lambda = 0.0	0.6320813
	Lambda = 0.5	0.3968958
	Lambda = 1.0	0.4908678
PAM	Threshold = 0.7608929	0.4986565
	Threshold = 11.0329476	1.9483062
	Threshold = 21.3050022	1.9483062
PLS	Ncomp = 1	1.826913
	Ncomp = 2	1.733439
	Ncomp = 3	1.643669
C5.0 tree		0.7482527
ET		0.3590714

KKNN: weighted k-nearest neighbors; PDA: penalized discriminant analysis; HDDA: high-dimensional discriminant analysis; SDA: shrinkage discriminant analysis; PAM: nearest shrunken centroids; PLS: partial least squares; ET: extremely randomized trees.

For the PDA algorithm, the optimal lambda value was 0.1. For HDDA, the best model had a threshold of 0.300, but this algorithm is not robust and other repetitions led to a different configuration; for SDA, the lowest log loss was obtained with lambda = 0.05, and for PAM, the best model had a threshold = 0.70615. This value can change slightly if the whole treatment is repeated. With PLS, log loss was also used to select the optimal model using the smallest value, and the final value selected for the model was as follows: number of components (ncomp) = 3 (Table 1).

The box plots for the three main metric parameters log loss, accuracy and kappa for eight classifiers can be observed in Figure 4.

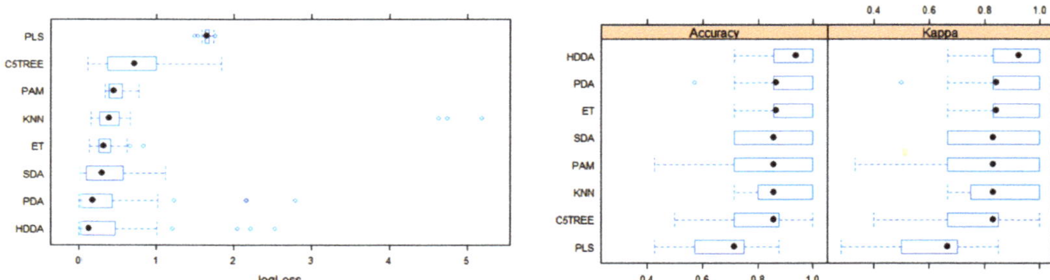

Figure 4. Box plots of log loss, accuracy and kappa values for various machine learning (ML) algorithms after training with 10-fold cross-validation to obtain the best model using the training dataset. Black circles symbolize mean values. PLS: partial least squares; C5TREE: C5.0 tree; PAM: nearest shrunken centroids; KNN: weighted k-nearest neighbors; ET: extremely randomized trees; SDA: shrinkage discriminant analysis; PDA: penalized discriminant analysis; HDDA: high-dimensional discriminant analysis.

ANN (single-layer perceptron) was applied to the training set with 10-fold cross-validation. The training process evaluated from 1 to 20 hidden units (neurons) and weight decays from 0.1 to 0.5. After optimization of tuning parameters to maximize the validation accuracy, the best model had 17 hidden units and weight decay = 0.1 (Figure 5). As it can be observed, the variability of accuracy with more than five hidden units is low, ranging from 0.85 to 0.91. This means that repetitions of the treatments can produce different topologies with very similar accuracy.

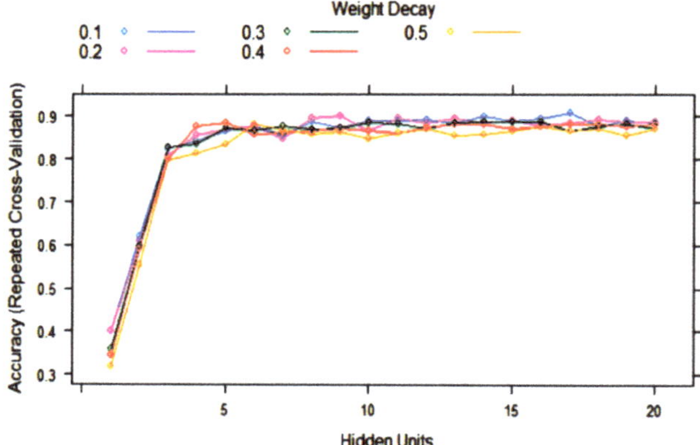

Figure 5. Change in the artificial neural network (ANN) accuracy during training with 10-fold cross-validation with the number of hidden units (nodes) and weight decay.

The accuracy of the SVM with linear kernels (SVM_L) algorithm during training with 10-fold cross-validation was maximized, with a value of the cost function of C = 2.5 (Figure 6). The largest value of the accuracy (0.61) was relatively low. In an attempt to improve SVM, we tested SVM with radial basis function kernels (SVM_R). The final values used for the SVM_R model were sigma = 0.1 and C = 0.5 (Figure 6). The accuracy was 0.562, meaning it was not improved. However, the cost value of these algorithms was quite variable on repeated treatments, maintaining the same partition ratio.

Figure 7 shows the variation in RF accuracy throughout training with 10-fold cross-validation. The final values for the RF model were as follows: number of variables randomly sampled as candidates at each split (mtry) = 8; the number of trees (ntry) parameter was 500; the maximum accuracy was 0.9246.

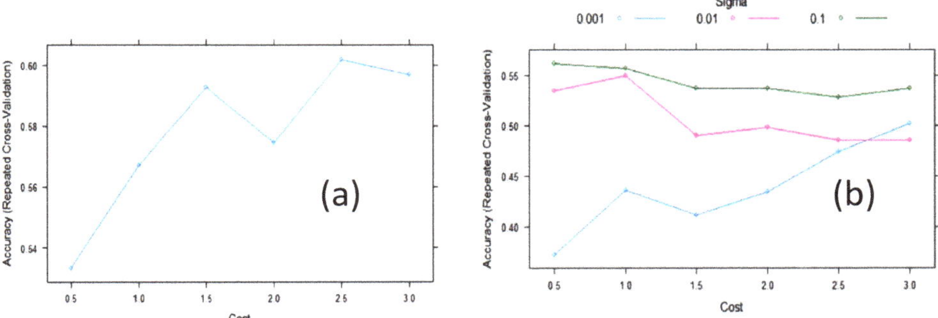

Figure 6. Change in accuracy of (**a**) Support vector machine with linear kernel (SVM_L) and (**b**) Support vector machine with radial kernel (SVM_R) during training with 10-fold cross-validation (CV) with the cost function.

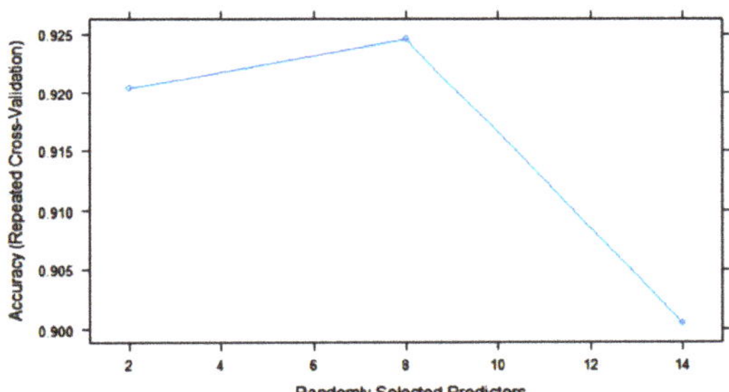

Figure 7. Change in the accuracy of random forest (RF) models with the number of randomly selected predictors (mtry).

The XGBoost tree algorithm has many parameters to tune, although, usually, some of them are held constant. Figure 8 shows the variation in some tuning parameters during the training process. The largest accuracy was obtained with "subsample" = 0.5, "shrinkage (eta)" = 0.1 and "max tree depth" = 4. Other final values for the model were "nrounds" = 200, "gamma" = 0, "colsample_bytree" = 0.8 and "min_child_weight" = 1.

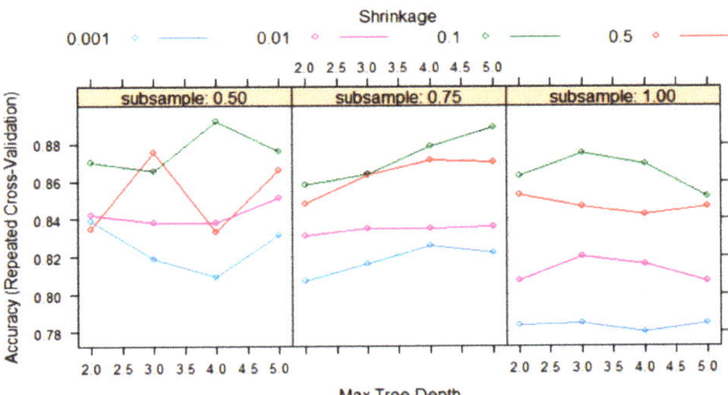

Figure 8. Change in the accuracy of XGBoost algorithm during training with 10-fold cross-validation with the parameters "max tree depth", "shrinkage (eta)" and "subsample". Tuning parameters "nrounds", "gamma", "colsample_bytree" and "min_child_weight" had constant values of 200, 0, 0.8 and 1, respectively.

The confusion matrices produced by all the ML models on the test set (30 samples) are listed in Tables 2–4. These matrices show the true botanical origin (Reference) in the columns and the predicted classification (Prediction by the models) in the rows. The ideal situation is to have all the samples located on the diagonal cells of the matrix, which would mean that the accuracy is 100%. The overall accuracies obtained with the PDA, SDA, ET, PLS and 5.0 tree algorithms were 90.00%, 86.67%, 86.67%, 73.33% and 76.67%, respectively (Table 2). The overall accuracies obtained with the KKNN, PAM, HDDA, ANN and RF algorithms were 83.33%, 83.33%, 83.33%, 86.67% and 80.00%, respectively (Table 3). The overall accuracies obtained with SVM with linear kernels (SVM_L), SVM with radial kernels (SVM_R) and XGBoost were 66.33%, 60.00% and 90.00%, respectively (Table 4).

Table 2. Confusion matrices of various classifier algorithms (PDA, SDA, ET, PLS, C5.0 tree) on the test set. The number of honey samples in this set was as follows: citrus (5), eucalyptus (4), forest (5), heather (4), lavender (4), rosemary (4) and sunflower (4).

		Reference						
	Prediction	**Citrus**	**Eucalyptus**	**Forest**	**Heather**	**Lavender**	**Rosemary**	**Sunflower**
PDA	Citrus	5	0	0	0	0	1	0
	Eucalyptus	0	4	0	0	1	0	0
	Forest	0	0	5	0	0	0	0
	Heather	0	0	0	4	0	0	0
	Lavender	0	0	0	0	2	0	0
	Rosemary	0	0	0	0	0	3	0
	Sunflower	0	0	0	0	1	0	4
SDA	Citrus	4	0	0	0	0	1	0
	Eucalyptus	0	4	0	0	1	0	0
	Forest	0	0	5	0	0	0	0
	Heather	0	0	0	4	0	0	0
	Lavender	0	0	0	0	3	1	0
	Rosemary	1	0	0	0	0	2	0
	Sunflower	0	0	0	0	0	0	4
ET	Citrus	4	0	0	0	0	0	0
	Eucalyptus	0	4	0	0	2	0	0
	Forest	0	0	5	0	0	0	0
	Heather	0	0	0	4	0	0	0
	Lavender	0	0	0	0	2	1	0
	Rosemary	1	0	0	0	0	3	0
	Sunflower	0	0	0	0	0	0	4
PLS	Citrus	5	0	0	0	0	3	0
	Eucalyptus	0	3	0	0	0	0	0
	Forest	0	0	5	0	0	1	0
	Heather	0	0	0	4	0	0	0
	Lavender	0	0	0	0	1	0	0
	Rosemary	0	0	0	0	0	0	0
	Sunflower	0	1	0	0	3	0	4
C5.0 tree	Citrus	4	1	0	0	0	1	0
	Eucalyptus	0	2	0	0	2	0	0
	Forest	0	0	5	0	0	0	0
	Heather	0	0	0	4	0	0	0
	Lavender	0	1	0	0	1	0	0
	Rosemary	0	0	0	0	0	3	0
	Sunflower	1	0	0	0	1	0	4

Table 3. Confusion matrices of various classifier algorithms (KKNN, PAM, HDDA, ANN and RF) on the test set. The number of honey samples was the same as that indicated in Table 2.

		Reference						
	Prediction	**Citrus**	**Eucalyptus**	**Forest**	**Heather**	**Lavender**	**Rosemary**	**Sunflower**
KKNN	Citrus	4	0	0	0	0	1	0
	Eucalyptus	0	4	0	0	0	0	0
	Forest	0	0	5	0	0	0	0
	Heather	0	0	0	4	0	0	0
	Lavender	0	0	0	0	1	0	0
	Rosemary	1	0	0	0	0	3	0
	Sunflower	0	0	0	0	3	0	4
PAM	Citrus	5	0	0	0	0	1	0
	Eucalyptus	0	3	0	0	1	0	0
	Forest	0	0	5	0	0	0	0
	Heather	0	0	0	4	0	0	0
	Lavender	0	0	0	0	2	0	0
	Rosemary	0	0	0	0	0	3	0
	Sunflower	0	1	0	0	1	0	4
HDDA	Citrus	4	0	0	0	0	1	0
	Eucalyptus	0	4	0	0	1	0	0
	Forest	0	0	4	0	0	0	0
	Heather	0	0	1	4	0	0	0
	Lavender	0	0	0	0	2	0	0
	Rosemary	0	0	0	0	0	3	0
	Sunflower	1	0	0	0	1	0	4
ANN	Citrus	5	0	0	0	0	1	0
	Eucalyptus	0	4	0	0	0	0	0
	Forest	0	0	5	0	0	0	0
	Heather	0	0	0	4	0	0	0
	Lavender	0	0	0	0	1	0	0
	Rosemary	0	0	0	0	0	3	0
	Sunflower	0	0	0	0	3	0	4
RF	Citrus	5	1	0	0	0	1	0
	Eucalyptus	0	2	0	0	1	0	0
	Forest	0	0	5	1	0	0	0
	Heather	0	0	0	3	0	0	0
	Lavender	0	1	0	0	2	0	0
	Rosemary	0	0	0	0	0	3	0
	Sunflower	0	0	0	0	1	0	4

An accuracy of 100% was not obtained with any model. The largest accuracy was provided by the models obtained with PDA and XGBoost (90%) followed by SDA, ET and ANN. The lowest accuracy was provided by SVM, especially SVM_R, which failed to correctly classify all heather, lavender and rosemary honeys. All models correctly classified sunflower honeys, and most of them (11) correctly classified all forest honeys. Ten models correctly classified the four heather honeys. Seven models correctly classified all citrus honeys. Only XGBoost classified the four rosemary honeys into their parent groups; no model was able to correctly classify all the lavender honeys. Some lavender honeys were classified as sunflower by XGBoost, SVM, ANN, PAM, RF, HDDA, KKNN, C5.0 tree, PLS and PDA. Other lavender honeys were classified as eucalyptus honeys.

To test the robustness of the overall accuracies, classifications were repeated three more times (the samples included in the training and test sets changed randomly) while maintaining the same splitting ratio (70/30). The box plots of the metrics (log loss, accuracy and kappa) of some optimized models are shown in Figure S5. The results of overall mean accuracies of all the models on the test sets after four repetitions of the whole process (training/10-fold cross-validation), including the ones shown above (Tables 2–4), are summarized in Table 5.

Table 4. Confusion matrices of various classifier algorithms (SVM_L, SVN_R and XGBoost) on the test set. The number of honey samples was the same as that indicated in Table 2.

		Reference						
	Prediction	**Citrus**	**Eucalyptus**	**Forest**	**Heather**	**Lavender**	**Rosemary**	**Sunflower**
SVM_L	Citrus	3	0	0	0	0	1	0
	Eucalyptus	0	4	0	1	1	0	0
	Forest	0	0	5	2	0	0	0
	Heather	0	0	0	0	0	0	0
	Lavender	0	0	0	1	0	0	0
	Rosemary	2	0	0	0	1	3	0
	Sunflower	0	0	0	0	2	0	4
SVM_R	Citrus	5	0	0	1	0	4	0
	Eucalyptus	0	4	0	0	2	0	0
	Forest	0	0	5	0	0	0	0
	Heather	0	0	0	0	0	0	0
	Lavender	0	0	0	3	0	0	0
	Rosemary	0	0	0	0	0	0	0
	Sunflower	0	0	0	0	2	0	4
XGB	Citrus	5	0	0	0	0	0	0
	Eucalyptus	0	4	0	0	0	0	0
	Forest	0	0	4	0	0	0	0
	Heather	0	0	1	4	0	0	0
	Lavender	0	0	0	0	2	0	0
	Rosemary	0	0	0	0	0	4	0
	Sunflower	0	0	0	0	2	0	4

Table 5. Overall accuracy of ML models for classification of honey samples in the test sets.

ML Algorithm	Overall Accuracy per Test				Mean Overall Accuracy
	Test 1	**Test 2**	**Test 3**	**Test 4**	
PLS	0.7333	0.6667	0.6333	0.7000	0.6833
C5.0 tree	0.7667	0.7667	0.7667	0.8000	0.7750
KKNN	0.8333	0.8333	0.7000	0.8000	0.7916
PAM	0.8333	0.8333	0.6667	0.8667	0.8000
PDA	0.9000	0.9333	0.7667	0.8667	0.8667
SDA	0.8667	0.8667	0.7667	0.8333	0.8333
ET	0.8333	0.8667	0.7667	0.9000	0.8417
HDDA	0.8333	0.8667	0.7667	0.9000	0.8417
ANN	0.8667	0.9333	0.7667	0.8667	0.8584
RF	0.8000	0.8333	0.8667	0.8667	0.8417
SVML	0.6333	0.4667	0.5000	0.6667	0.5667
SVMR	0.6000	0.6667	0.5333	0.5667	0.5917
XGBoost	0.9000	0.8333	0.7000	0.9333	0.8417

As deduced from the results in Table 5, the PDA algorithm had the largest mean overall accuracy on the test set (86.67%), followed by ANN (85.84%), ET, RF and XGBoost (84.17%). The worst performance was rendered by SVM_L and SVM_R ($\leq$60%). The most stable algorithm was C5.0 tree.

In the case that all samples are used for training with 10-fold cross-validation without separation of a test set, the results are much better with all the models. The training was performed similar to the case of splitting, using the same parameters (log loss, accuracy, kappa) for obtaining the best models (Figure S6). This approach is sometimes found in the literature concerning honey classification, but overfitting is usually a problem. The overall accuracies in this case, according to the confusion matrices (Table S2), were 100% for ET, RF and XGBoost, 97% for PDA and ANN, 95% for C5.0 tree, 92% for SDA, 91% for PAM, 90% for KKNN, 87% for HDDA, 69% for PLS, 66% for SVM_R and 56% for SVM_L.

4. Discussion

A variety of factors can influence the variability of the data observed in the dataset. For example, early harvest to increase the amount of honey especially of citrus or rosemary may lead to unripe, very clear products that do not meet legal requirements, although they can be unifloral. Beekeepers or traders can store these crops or blend early with late crops to obtain acceptable products. Early harvest can also affect the sugar content because unripe citrus honey may have more than 20% sucrose. In this case, they cannot be placed directly on the market, although they may be blended with more ripened honeys of the same class to comply with regulations. Late harvest may affect these variables as a large amount of pollen and more variability in the pollen spectrum are expected to occur because bees will go on gathering all available flowers or honeydew and pollen. Therefore, it may be very difficult to obtain unifloral honeys. The more time honey remains inside the combs, the riper the honey is expected to be, with a very low amount of sucrose; the contents of fructose and glucose may vary at low levels or increase, except in the case of honeydew honey, and a safe level of moisture will be reached. Thus, a good balance between early and late harvest should be taken into account by beekeepers to obtain unifloral honeys with the best quality.

In the present study, different ML algorithms using R and the caret package were applied to the same dataset of honeys belonging to seven classes from a single botanical origin collected in Spain. The initial dataset was a matrix of 100 honey samples (rosemary 13, citrus 16, lavender 14, eucalyptus 14, heather 13, sunflower 14 and forest 16) and 14 physiochemical features (water content, electrical conductivity, pH, sugars and colorimetric coordinates). It was partitioned into a training set and a test set.

The first approach to analyze the dataset considered an unsupervised approach. The data were analyzed by PCA and k-means, and two broad clusters with 70 and 30 samples were shown using k-means clustering. The two clusters are too broad to meet the actual honey classes. All variables were used because all were considered important by Boruta. Then, supervised ML approaches were tested. ML classifier algorithms applied were KKNN, PDA, PLS, PAM, HDDA, SDA, C5.0tree, ET, ANN, SVM_L, SVM_R, RF and XGBoost.

The metric used for optimizing most models was log loss. Other metric parameters such as accuracy or kappa were also calculated and used instead of log loss for ANN, RF or XGBoost. After training using the same randomly selected dataset (70 samples) and finding the optimal configuration using 10-fold cross-validation, the performance of all models to accurately classify the test samples into their parent classes was compared. All treatments were repeated four times under the same conditions although the samples were randomly distributed in both sets. The performance was not constant among repetitions, and the mean accuracy was considered. The best results were provided by the PDA classifier, which classified the unifloral honeys in the test set within their parent types with 86.67% overall accuracy on average. Good results were also obtained with ANN, ET, HDDA, RF and XGBoost, while SVML and SVMR proved to be the worst. The honeys that have the best chance to be correctly classified are sunflower, forest, heather, eucalyptus and citrus. The correct classification of rosemary honey was hard to carry out, but the most difficult to be appropriately classified were lavender honeys. A low number of samples in the test set can be a problem in making good predictions, especially with samples that have a low percentage of pollen of the putative taxa or have pollen from other nectariferous plants. This happens with lavender, rosemary and citrus honeys, with under-represented pollen [13], and is a problem because they are very appreciated by consumers [60]. It has been reported that pollen analysis can be of limited usefulness for labeling lavender honeys, and analysis of volatiles should be considered as a complementary technique in the case that samples show the characteristic organoleptic properties [21,61].

Application of supervised ML algorithms to the classification of unifloral honeys by botanical origin is an issue of interest. This classification is based on the labeling of the studied samples into classes based on melissopalynology and organoleptic properties. However, microscopic analysis is time-consuming, requires highly specialized personnel and is unable to detect seasonal variation in pollen amounts or fraudulent pollen.

Former attempts at classification were conducted using multivariate statistical discriminant techniques applied to physicochemical features [11], with rather good results. Anjos et al. [36] investigated different ANN configurations to classify the botanical origin of 49 honey samples. Measurements of moisture, electrical conductivity, water activity, ash content, pH, free acidity, colorimetric coordinates and total phenol content were used as input variables. It was concluded that the botanical origin of honey can be reliably and quickly known from the colorimetric information and the electrical conductivity of the honey, which agrees with our results. Another report [24] showed the results obtained with a similar set of variables, although including a large phenolic profile, to classify acacia, tilia (linden), sunflower, honeydew and polyfloral honeys of Romanian origin (50 samples) labeled by pollen analysis into their parent classes by using LDA and ANN as classifiers. LDA correctly classified 92.0% of the samples. An ANN with two hidden layers classified 94.8% of the honey samples into their botanical origin. However, all samples from each class were used to reach these accuracy rates. In the present paper, a test set was used to calculate the percentage of correctly assigned origins, and we obtained higher accuracy rates using all the samples. Popek et al. [37] were able to correctly classify nearly all their samples according to their botanical origin using CART. They obtained good results using all 72 samples (9 samples $\times$ 8 classes) under treatment (rape, acacia, heather, linden, buckwheat, honeydew, nectar-honeydew and multifloral honeys).

Authentication of honey origin using ML algorithms and nondestructive analytical techniques has been reported. In this way, ATR-FTIR spectra of 130 Serbian samples belonging to acacia, linden and sunflower honeys were treated by SVM, and the predictability rate was high [27], although the classes only totaled three, and the method of carrying out sample labeling was omitted. In our treatment, SVM was not useful. Ciulu et al. [62] reported on the usage of ATR-FTIR spectra and processing by RF to this aim. Eighty samples belonging to four different floral origins were considered: strawberry tree, asphodel, thistle and eucalyptus. Training an RF on the IR spectra allowed achieving an average accuracy of 87% in a cross-validation setting. This is approximately the same accuracy rate obtained in our study using different variables. FT-Raman spectra combined with PCA or LDA have also proved to be useful to classify monofloral honeys with a high degree of accuracy [29–31].

Using NIR (850–2500 nm), classification of 119 Italian honey samples encompassing acacia, linden and chestnut unifloral honeys and multifloral honeys was attempted by PLS and SVM with linear kernels using cross-validation of the NIR spectra [63]. Pollen analysis was not used for labeling. SVM provided better classification scores than PLS contrary to what happens in our case. An additional approach was to apply Boruta for feature selection, but the accuracy was not improved. Splitting of the dataset into a training/CV set and an independent test set was not carried out, meaning that confusion matrices included all samples, which obviously improves the success rate. Linden honeys failed to be correctly classified, which might be due to the low number of samples of that class. NIR was also the source of input variables to classify five types of Chinese unifloral honeys by application of Mahalanobis distance discriminant analysis (MD-DA) and a backpropagation artificial neural network (BP-ANN) [64]. By the MD-DA model, overall correct classification rates were 87.4% and 85.3% for the calibration and validation samples, respectively, while the ANN model resulted in having total correct classification rates of 90.9% and 89.3% for the calibration and validation sets, respectively. Pollen analysis was not employed for origin assignation to honeys. Minaei et al. [28] used VIS–NIR hyperspectral images of 52 samples of five classes of unifloral honeys and, after a reduction in dimensionality, applied RBF networks (a type of ANN with several distinctive features), RF and SVM for classification.

The test set had 20 samples, and the remaining 32 samples were used for training. The first ML rendered 92% accuracy, while SVM and RF returned accuracies of 84 and 89%, respectively. A problem related to this technique is the variability in color with time.

Other types of input variables within the group of nondestructive methodologies are based on sensors able to mimic organoleptic perceptions such as the electronic nose (E-nose) [34,65] and the electronic tongue [35]. The E-nose generates signals corresponding to volatile and semivolatile compounds from honeys that, after being processed by ML algorithms, have the ability to carry out correct classifications. Benedetti et al. [65] studied 70 samples ascribed to three unifloral origins, which were certificated by pollen analysis. First, a PCA of samples indicated the main components, and then an ANN was generated that, after optimization by cross-validation, was able to accurately classify all samples of the test set. An electronic tongue was reported to correctly classify acacia, chestnut and honeydew honeys after application of an ANN to signals from the device [35].

Thus, the application of ML algorithms to classification of unifloral honeys has been increasing in recent years, and it is expected that it will go on increasing in the future. However, a systematic comparison of the main ML algorithms to reach this goal as it is presented in this study has not been reported to date. The PDA algorithm was the best, but others such as ANN, SDA, RF, ET or HDDA can also be useful to perform accurate classifications based on the variables from the dataset. SVM worked badly with all repetitions on the datasets. Failure in obtaining larger accuracy rates is due to some honey classes such as lavender, rosemary or citrus with under-represented pollen grains. Good marker parameters should be found and used to improve the classification of these honeys that have not been included in most studies using ML algorithms for prediction. To our knowledge, this is the first time that most of the compared algorithms in the present study (for example, PDA, HDDA, SDA, C5.0, ET, XGBoost) have been used for the goal of classification of unifloral honeys. It is expected that the comparison of the performance of the ML algorithms applied here may be useful not only for research on the topic of honey classification by origin but also for research on other kinds of foods.

5. Conclusions

A comparison of 13 ML algorithms on a dataset of one hundred honeys harvested in Spain and belonging to seven unifloral classes was performed using 14 physicochemical parameters. The ML algorithms were built by splitting the dataset into a training set (70%) and a test set (30%) and optimizing the configuration by 10-fold cross-validation using several parameters, but mainly log loss. The optimized models were tested on the test set to record the overall and partial accuracies in the right classification of samples into their parent classes. The whole process was repeated three times, and the results were averaged. The best accuracies were provided by the PDA algorithm, (86.67%), followed by ANN (85.56%), SDA and RF (83.33%). The worst results were rendered by SVM with radial and linear kernels (53–60%). Most algorithms correctly classified forest, sunflower and heather honeys. Orange blossom and eucalyptus honey samples were partly misclassified by some models; rosemary honeys were partly misclassified by all models, except XGBoost, while lavender honeys were the most difficult to be included into their parent groups. Most the algorithms studied here have not been applied previously to the issue of honey classification, and they can likely be useful for such a task in future research such as the inclusion of more unifloral honey types and a multifloral honey class. Moreover, other parameters (among them those obtained by FT-IR, FT-Raman or NIR spectroscopy nondestructive techniques) can be included in the datasets and tested to improve the accuracy of the classification task as much as possible.

Supplementary Materials: The following are available online at https://www.mdpi.com/article/10.3390/foods10071543/s1, Figure S1: Box plots of some variables in the honey dataset: (a) water content (%); (b) electrical conductivity (µs/cm); (c) pH; (d) fructose content (%); (e) glucose content (%); (f) sucrose content (%); (g) maltose content (%); (h) isomaltose content (%), Figure S2: Box plots of some variables in the honey dataset: (a) kojibiose content (%); (b) fructose/glucose ratio; (c) glucose/water ratio; (d) chromatic coordinate x; (e) chromatic coordinate y; (f) chromatic coordinate L; (g) percentage of pollen from the characteristic taxa giving the name to each unifloral honey (this variable is not included in the dataset), Figure S3: Clustering of the honey dataset in 7 clusters by k-means. The largest symbols correspond to the centroids of each cluster, Figure S4: Plot of the importance of the variables from the honey dataset, as estimated by an RF model, Figure S5: Box plots of metric values of several ML algorithms by optimization throughout training with 10-fold cross-validation of the honey dataset with splitting into training and test sets obtained in three additional repetitions: (a), (b) and (c) log loss of repetitions 2, 3 and 4; (d), (e) and (f) accuracy and kappa of repetitions 2, 3 and 4. Black circles symbolize mean values, Figure S6: Box plots of (a) log loss, (b) accuracy and kappa values of several ML algorithms throughout the training with 10-fold cross-validation of the honey dataset without partitioning into training and test sets. Black circles symbolize mean values, Table S1: Mean values for the predictor variables in the two clusters obtained by k-means, Table S2: Confusion matrices obtained with all honeys in the dataset without splitting into training and test sets using 10-fold cross-validation.

Author Contributions: Conceptualization, F.M. and E.M.M.; methodology, F.M., E.M.M. and A.T.; software, F.M.; validation, F.M.; formal analysis, F.M.; investigation, F.M., E.M.M. and A.T.; resources, F.M.; data curation, F.M.; writing—original draft preparation, F.M.; writing—review and editing, F.M. and E.M.M.; visualization, F.M.; supervision, F.M. and E.M.M.; project administration, E.M.M.; funding acquisition, E.M.M. All authors have read and agreed to the published version of the manuscript.

Funding: This research received co-funding from EUROPEAN REGIONAL DEVELOPMENT FUND (ERDF) and MINISTERIO DE ECONOMÍA Y COMPETITIVIDAD (Spanish Government) through project RTI2018-097593-B-C22.

Data Availability Statement: The data supporting the results can be found in Supplementary Materials or by petition to the corresponding author.

Conflicts of Interest: The authors declare no conflict of interest.

References

1. Cianciosi, D.; Forbes-Hernández, T.Y.; Afrin, S.; Gasparrini, M.; Reboredo-Rodriguez, P.; Manna, P.P.; Zhang, J.; Bravo Lamas, L.; Martínez Flórez, S.; Agudo Toyos, P.; et al. Phenolic compounds in honey and their associated health benefits: A review. *Molecules* **2018**, *23*, 2322. [CrossRef] [PubMed]
2. Afrin, S.; Haneefa, S.M.; Fernandez-Cabezudo, M.J.; Giampieri, F.; al-Ramadi, B.K.; Battino, M. Therapeutic and preventive properties of honey and its bioactive compounds in cancer: An evidence-based review. *Nutr. Res. Rev.* **2020**, *33*, 50–76. [CrossRef] [PubMed]
3. European Commission. Regulation (EC) No 178/2002 of the European Parliament and of the council of 28 January 2002 laying down the general principles and requirements of food law, establishing the European food safety authority and laying down procedures in matters of food safety. *Off. J. Eur. Commun.* **2002**, *L 31*, 1–24.
4. Council Directive 2001/110/EC of 20 December 2001 relating to honey. *Off. J. Eur. Comm.* **2001**, *L 10*, 47–52.
5. Directive 2014/63/EU of the European Parliament and of the Council of 15 May 2014 amending Council Directive 2001/110/EC relating to honey. *Off. J. Eur. Union* **2014**, *L 164*, 1–5.
6. Codex Alimentarius Standard for honey CXS 12-1981 Adopted in 1981. Revised in 1987, 2001. Amended in 2019. 2001. Available online: http://www.fao.org/fao-who-codexalimentarius/sh-proxy/en/?lnk=1&url=https%253A%252F%252Fworkspace.fao.org%252Fsites%252Fcodex%252FStandards%252FCXS%2B12-1981%252FCXS_012e.pdf (accessed on 21 June 2021).
7. Ampuero, S.; Bogdanov, S.; Bosset, J.O. Classification of unifloral honeys with an MS-based electronic nose using different sampling modes: SHS, SPME and INDEX. *Eur. Food Res. Technol.* **2004**, *218*, 198–207. [CrossRef]
8. Cavaco, A.M.; Miguel, G.; Antunes, D.; Guerra, R. Determination of geographical and botanical origin of honey: From sensory evaluation to the state of the art of non-invasive technology. In *Honey: Production, Consumption and Health Benefits*; Bondurand, G., Bosch, H., Eds.; Nova Science Publishers: Hauppauge, NY, USA, 2012; pp. 1–40.
9. Maurizio, A. Microscopy of honey. In *Honey: A Comprehensive Survey*; Crane, E., Ed.; Heinemann in Cooperation with the International Bee Research Association: London, UK, 1975; pp. 240–257.
10. Louveaux, J.; Maurizio, A.; Vorwohl, G. Methods of melissopalynology. *Bee World* **1978**, *59*, 139–157. [CrossRef]
11. Mateo, R.; Bosch-Reig, F. Classification of Spanish unifloral honeys by discriminant analysis of electrical conductivity, color, water content, sugars, and pH. *J. Agric. Food Chem.* **1998**, *46*, 393–400. [CrossRef] [PubMed]
12. White, J.W.; Bryant, V.M., Jr. Assessing citrus honey quality: Pollen and methyl anthranilate content. *J. Agric. Food Chem.* **1996**, *44*, 3423–3425. [CrossRef]
13. Persano-Oddo, L.; Piro, R. Main European unifloral honeys: Descriptive sheets1. *Apidologie* **2004**, *35*, S38–S81. [CrossRef]

14. Mateo Castro, R.; Jiménez Escamilla, M.; Bosch Reig, F. Evaluation of the color of some Spanish unifloral honey types as a characterization parameter. *J. AOAC Int.* **1992**, *75*, 537–542. [CrossRef]
15. Mateo, R.; Bosch-Reig, F. Sugar profiles of Spanish unifloral honeys. *Food Chem.* **1997**, *60*, 33–41. [CrossRef]
16. de la Fuente, E.; Ruiz-Matute, A.I.; Valencia-Barrera, R.M.; Sanz, J.; Martínez Castro, I. Carbohydrate composition of Spanish unifloral honeys. *Food Chem.* **2011**, *129*, 1483–1489. [CrossRef]
17. Weston, R.J.; Brocklebank, L.K. The oligosaccharide composition of some New Zealand honeys. *Food Chem.* **1999**, *64*, 33–37. [CrossRef]
18. Bouseta, A.; Scheirman, V.; Collin, S. Flavor and free amino acid composition of lavender and eucalyptus honeys. *J. Food Sci.* **1996**, *61*, 683–687, 694. [CrossRef]
19. Baroni, M.V.; Nores, M.L.; Díaz, M.D.P.; Chiabrando, G.A.; Fassano, J.P.; Costa, C.; Wunderlin, D.A. Determination of volatile organic compound patterns characteristic of five unifloral honey by solid-phase microextraction−gas chromatography−mass spectrometry coupled to chemometrics. *J. Agric. Food Chem.* **2006**, *54*, 7235–7241. [CrossRef]
20. Revell, L.E.; Morris, B.; Manley-Harris, M. Analysis of volatile compounds in New Zealand unifloral honeys by SPME–GC–MS and chemometric-based classification of floral source. *Food Meas.* **2014**, *8*, 81–91. [CrossRef]
21. Castro-Vázquez, L.; Díaz-Maroto, M.C.; González-Viñas, M.A.; Pérez-Coello, M.S. Differentiation of monofloral citrus, rosemary, eucalyptus, lavender, thyme and heather honeys based on volatile composition and sensory descriptive analysis. *Food Chem.* **2009**, *112*, 1022–1030. [CrossRef]
22. Machado, A.M.; Miguel, M.G.; Vilas-Boas, M.; Figueiredo, A.C. Honey volatiles as a fingerprint for botanical origin—A review on their occurrence on monofloral honeys. *Molecules* **2020**, *25*, 374. [CrossRef]
23. Sun, Z.; Zhao, L.; Cheng, N.; Xue, X.; Wu, L.; Zheng, J.; Cao, W. Identification of botanical origin of Chinese unifloral honeys by free amino acid profiles and chemometric methods. *J. Pharm. Anal.* **2017**, *7*, 317–323. [CrossRef]
24. Oroian, M.; Sorina, R. Honey authentication based on physicochemical parameters and phenolic compounds. *Comput Electron. Agric.* **2017**, *138*, 148–156. [CrossRef]
25. Karabagias, I.K.; Louppis, A.P.; Kontakos, S.; Drouza, C.; Papastephanou, C. Characterization and botanical differentiation of monofloral and multifloral honeys produced in Cyprus, Greece, and Egypt using physicochemical parameter analysis and mineral content in conjunction with supervised statistical techniques. *J. Anal. Meth. Chem.* **2018**, *7698251*. [CrossRef] [PubMed]
26. Ruoff, K.; Luginbühl, W.; Kilchenmann, V.; Bosset, J.O.; von der Ohec, K.; von der Ohec, W.; Amad, R. Authentication of the botanical origin of honey using profiles of classical measurands and discriminant analysis. *Apidologie* **2007**, *38*, 438–452. [CrossRef]
27. Lenhardt, L.; Zeković, I.; Dramićanin, T.; Tešić, Ž.; Milojković-Opsenica, D.; Dramićanin, M.D. Authentication of the botanical origin of unifloral honey by infrared spectroscopy coupled with support vector machine algorithm. *Phys. Scr.* **2014**, *T162*, 014042. [CrossRef]
28. Minaei, S.; Shafiee, S.; Polder, G.; Moghadam-Charkari, N.; van Ruth, S.; Barzegar, M.; Zahiri, J.; Alewijn, M.; Kus, P.M. VIS/NIR imaging application for honey floral origin determination. *Infrared Phys. Technol.* **2017**, *86*, 218–225. [CrossRef]
29. Corvucci, F.; Nobili, L.; Melucci, D.; Grillenzoni, F.V. The discrimination of honey origin using melissopalynology and Raman spectroscopy techniques coupled with multivariate analysis. *Food Chem.* **2015**, *169*, 297–304. [CrossRef]
30. Oroian, M.; Ropciuc, S. Botanical authentication of honeys based on Raman spectra. *Food Meas.* **2018**, *12*, 545–554. [CrossRef]
31. Xagoraris, M.; Lazarou, E.; Kaparakou, E.H.; Alissandrakis, E.; Tarantilisa, P.A.; Pappas, C.S. Botanical origin discrimination of Greek honeys: Physicochemical parameters versus Raman spectroscopy. *J. Sci. Food Agric.* **2021**, *101*, 3319–3327. [CrossRef]
32. Melucci, D.; Cocchi, M.; Corvucci, F.; Boi, M.; Tositti, L.; de Laurentiis, F.; Zappi, A.; Locatelli, C.; Locatelli, M. Chemometrics for the direct analysis of solid samples by spectroscopic and chromatographic techniques. In *Chemometrics: Methods, Applications and New Research*; Luna, A.S., Ed.; Nova Science Publishers: Hauppauge, NY, USA, 2017; pp. 173–204.
33. Siddiqui, A.J.; Musharraf, S.G.; Choudhary, M.I. Application of analytical methods in authentication and adulteration of honey. *Food Chem.* **2017**, *217*, 687–698. [CrossRef]
34. Zahed, N.; Najib, M.S.; Tajuddin, S.N. Categorization of gelam, acacia and tualang honey odor-profile using k-nearest neighbors. *Int. J. Soft. Eng. Comput Syst.* **2018**, *4*, 15–28. [CrossRef]
35. Major, N.; Marković, K.; Krpan, M.; Šarić, G.; Hrus˘kar, M.; Vahčić, N. Rapid honey characterization and botanical classification by an electronic tongue. *Talanta* **2011**, *85*, 569–574. [CrossRef]
36. Anjos, O.; Iglesias, C.; Peres, F.; Martínez, J.; García, Á.; Taboada, J. Neural networks applied to discriminate botanical origin of honeys. *Food Chem.* **2015**, *175*, 128–136. [CrossRef]
37. Popek, S.; Halagarda, M.; Kursa, K. A new model to identify botanical origin of Polish honeys based on the physicochemical parameters and chemometric analysis. *LWT Food Sci. Technol.* **2017**, *77*, 482–487. [CrossRef]
38. Maione, C.; Barbosa, F., Jr.; Barbosa, R.M. Predicting the botanical and geographical origin of honey with multivariate data analysis and machine learning techniques: A review. *Comput. Electron. Agric.* **2019**, *157*, 436–446. [CrossRef]
39. Escuredo, O.; Fernández-González, M.; Seijo, M.C. Differentiation of blossom honey and honeydew honey from Northwest Spain. *Agriculture* **2012**, *2*, 25–37.
40. Seijo, M.C.; Escuredo, O.; Rodríguez-Flores, M.S. Physicochemical properties and pollen profile of oak honeydew and evergreen oak honeydew honeys from Spain: A comparative study. *Foods* **2019**, *8*, 126. [CrossRef]

41. Orden de 12 de junio de 1986 por la que se aprueban los métodos oficiales de análisis para la miel. (Order of 12 June 1986 approving the official methods of analysis for honey). *BOE* **1986**, *145*, 22195–22202. Available online: https://www.boe.es/eli/es/o/1986/06/12/(3)/dof/spa/pdf (accessed on 22 June 2021).
42. AOAC 969. 38B MAFF Validated method V21 for moisture in honey. *J. Assoc. Public Anal.* **1992**, *28*, 183–187.
43. CIE (Commission Internationale de l'Eclairage). In Proceedings of the Eighth Session, Cambridge, UK, September 1931. Available online: http://classify.oclc.org/classify2/ClassifyDemo?owi=25128274 (accessed on 22 June 2021).
44. Hastie, T.; Bujas, A.; Tibsihrani, R. Penalized discriminant analysis. *Ann. Stat.* **1995**, *23*, 73–102. [CrossRef]
45. Hechenbichler, K.; Schliep, K. *Weighted k-Nearest-Neighbor Techniques and Ordinal Classification*; Sonderforschungsbereich 386, Paper 399; Ludwig-Maximilians-Universität: München, Germany, 2004; pp. 1–16. Available online: https://epub.ub.uni-muenchen.de/1769/1/paper_399.pdf (accessed on 4 May 2021).
46. Kuhn, M.; Wing, J.; Weston, S.; Williams, A.; Keefer, C.; Engelhardt, A.; Cooper, T.; Mayer, Z.; Kenkel, B.; The R Core Team; et al. Classification and Regression Training. R Package Version 2016, 6.0–71. Available online: https://CRAN.R-project.org/package=caret (accessed on 5 April 2021).
47. Bouveyron, C.; Girard, S.; Schmid, C. High-dimensional discriminant analysis. *Comm. Stat. Theor. Meth.* **2007**, *36*, 2607–2623. [CrossRef]
48. Tibshirani, R.; Hastie, T.; Narasimhan, B.; Chu, G. Diagnosis of multiple cancer types by shrunken centroids of gene expression. *Proc. Natl. Acad. Sci. USA* **2002**, *99*, 6567–6572. [CrossRef]
49. Tibshirani, R.; Hastie, T.; Narasimhan, B.; Chu, G. Class prediction by nearest shrunken centroids, with applications to DNA microarrays. *Stat. Sci.* **2003**, *18*, 104–117. [CrossRef]
50. Geurst, P.; Louis, D.; Wehenke, L. Extremely randomized trees. *Mach. Learn.* **2006**, *63*, 3–42.
51. Ahdesmäki, M.; Strimmer, K. Feature selection in omics prediction problems using cat scores and false non discovery rate control. *Ann. Appl. Stat.* **2010**, *4*, 503–519. [CrossRef]
52. Günther, F.; Fritsch, S. Neuralnet: Training of neural networks. *R J* **2010**, *2*, 30–38. [CrossRef]
53. Cortes, C.; Vapnik, V. Support-vector networks. *Mach. Learn.* **1995**, *20*, 273–297. [CrossRef]
54. Vapnik, V.N. *Statistical Learning Theory*; Wiley: New York, NY, USA, 1998.
55. Liaw, A.; Wiener, M.C. Classification and regression by random forest. *R News* **2002**, *2*, 18–22.
56. Chen, T.; Guestrin, C. XGBoost: A scalable tree boosting system. In Proceedings of the 22nd ACM SIGKDD International Conference on knowledge discovery and data mining, San Francisco, CA, USA, 13–17 August 2016; pp. 785–794. [CrossRef]
57. Bishop, C.M. *Neural Networks for Pattern Recognition*; Oxford University Press: Oxford, UK, 1995.
58. Tapas Kanungo, D.M. A local search approximation algorithm for k-means clustering. In Proceedings of the 18th Annual Symposium On Computational Geometry, Barcelona, Spain, 5–7 June 2002; ACM Press: New York, NY, USA, 2002; pp. 10–18.
59. Kursa, M.B.; Rudnicki, W.R. Feature selection with the boruta package. *J. Stat. Soft.* **2010**, *36*, 1–13. Available online: http://www.jstatsoft.org/v36/i11/ (accessed on 12 May 2021). [CrossRef]
60. Stevinho, L.M.; Chambó, E.D.; Pereira, A.P.R.; Carvalho, C.A.L.D.; de Toledo, V.D.A.A. Characterization of *Lavandula* spp. honey using multivariate techniques. *PLoS ONE* **2016**, *11*, e016220.
61. Escriche, I.; Sobrino-Gregorio, L.; Conchado, A.; Juan-Borrás, M. Volatile profile in the accurate labelling of monofloral honey. The case of lavender and thyme honey. *Food Chem.* **2017**, *226*, 61–68. [CrossRef]
62. Ciulu, M.; Oertel, E.; Serra, R.; Farre, R.; Spano, N.; Caredda, M.; Malfatti, L.; Sanna, G. Classification of unifloral honeys from SARDINIA (Italy) by ATR-FTIR spectroscopy and random forest. *Molecules* **2021**, *26*, 88. [CrossRef]
63. Bisutti, V.; Merlanti, R.; Serva, L.; Lucatello, L.; Mirisola, M.; Balzan, S.; Tenti, S.; Fontana, F.; Trevisan, G.; Montanucci, L.; et al. Multivariate and machine learning approaches for honey botanical origin authentication using near infrared spectroscopy. *J. Near Infrared Spectrosc.* **2019**, *27*, 65–74. [CrossRef]
64. Chen, L.; Wang, J.; Ye, Z.; Zhao, J.; Xue, X.; Vander Heyden, Y.; Sun, Q. Classification of Chinese honeys according to their floral origin by near infrared spectroscopy. *Food Chem.* **2012**, *135*, 338–342. [CrossRef] [PubMed]
65. Benedetti, S.; Mannino, S.; Sabatini, A.G.; Marcazzan, G.L. Electronic nose and neural network use for the classification of honey. *Apidologie* **2004**, *35*, 397–402. [CrossRef]

MDPI

Article

Sensory Descriptor Analysis of Whisky Lexicons through the Use of Deep Learning

Chreston Miller [1,*], Leah Hamilton [2] and Jacob Lahne [2]

[1] Data Services, University Libraries, Virginia Tech, 560 Drillfield Dr., Blacksburg, VA 24061, USA
[2] Department of Food Science & Technology, Virginia Tech, Blacksburg, VA 24061, USA; hleah@vt.edu (L.H.); jlahne@vt.edu (J.L.)
* Correspondence: chmille3@vt.edu

Abstract: This paper is concerned with extracting relevant terms from a text corpus on whisk(e)y. "Relevant" terms are usually contextually defined in their domain of use. Arguably, every domain has a specialized vocabulary used for describing things. For example, the field of Sensory Science, a sub-field of Food Science, investigates human responses to food products and differentiates "descriptive" terms for flavors from "ordinary", non-descriptive language. Within the field, descriptors are generated through Descriptive Analysis, a method wherein a human panel of experts tastes multiple food products and defines descriptors. This process is both time-consuming and expensive. However, one could leverage existing data to identify and build a flavor language automatically. For example, there are thousands of professional and semi-professional reviews of whisk(e)y published on the internet, providing abundant descriptors interspersed with non-descriptive language. The aim, then, is to be able to automatically identify descriptive terms in unstructured reviews for later use in product flavor characterization. We created two systems to perform this task. The first is an interactive visual tool that can be used to tag examples of descriptive terms from thousands of whisky reviews. This creates a training dataset that we use to perform transfer learning using GloVe word embeddings and a Long Short-Term Memory deep learning model architecture. The result is a model that can accurately identify descriptors within a corpus of whisky review texts with a train/test accuracy of 99% and precision, recall, and F1-scores of 0.99. We tested for overfitting by comparing the training and validation loss for divergence. Our results show that the language structure for descriptive terms can be programmatically learned.

Keywords: natural language processing; deep learning; sensory science; flavor lexicon; long short-term memory

Citation: Miller, C.; Hamilton, L.; Lahne, J. Sensory Descriptor Analysis of Whisky Lexicons through the Use of Deep Learning. *Foods* **2021**, *10*, 1633. https://doi.org/10.3390/foods10071633

Academic Editor: Sigfredo Fuentes

Received: 8 June 2021
Accepted: 7 July 2021
Published: 14 July 2021

1. Introduction

1.1. Flavor Language

Flavor is a major factor motivating eating behavior and food choice, but due to the approximately 350 different receptors for aroma-active compounds and the low detection thresholds for many such compounds [1], it is not currently possible to predict a flavor experience from chemical data alone. On the other hand, English and most other languages do not have a systematic and unambiguous flavor vocabulary, so studying flavor by surveying humans is still challenging. Sensory scientists need a way of aligning the different sensory lexicons used by different tasters and stakeholders.

The earliest solution in Sensory Science was the Descriptive Analysis (DA) panel, a body of related methods that use the experiences and vocabularies of a small panel of participants to create a single aligned sensory lexicon for some category of products. In order to ensure alignment between panelists (namely preventing needless synonyms and disagreement about definitions), every word is defined in reference to a physical standard. The largest time investment (often taking weeks or months) in DA is the hands-on training to identify appropriate descriptors and references, then creating experts in the newly-defined language, all of which occurs before the product analysis [2].

This standardized vocabulary is called a sensory or descriptive lexicon and comprises words or phrases called "descriptors": terms that can be used to describe the flavor, aroma, mouthfeel, taste, appearance, or other sensory attributes of the product set [3]. For highly-studied categories or in cases where flavor communication between groups is important, the lexicon itself may be a desired outcome [3,4]. Lexicons can provide a reference list of possible terms for analysis of products by trained or untrained panelists; lexicons can be used to communicate sensory properties of products to consumers for marketing and product differentiation, and lexicons can be used to define product categories by connoisseurs and enthusiasts, as prototypically documented in the wine world [3–5].

Smaller research operations and newer or less-studied categories, however, often prefer methods of flavor measurement that do not need a carefully crafted flavor lexicon. These "rapid" methods collect similarity measurements, allow consumers to use their own, untrained vocabulary to describe product flavors or both [6]. When flavor is described with colloquial language, there are likely to be individual differences and other problems like those encountered during DA training, but rapid methods deal with these problems after data collection rather than before.

The process of identifying meaningful descriptors from free-text product descriptions and combining synonymous terms is known as comment analysis (or free text analysis), and its adoption within Sensory Science has made it possible to utilize existing sources of descriptive text as sensory data. Comment analysis of existing text data has previously been used to produce or modify lexicons for rum [7], wine [8], and whisky [9] and to identify terms that drive liking or price in wine [10,11]. Like all descriptive lexicons, these will never be truly universal—a new product can have a new taste, or a new consumer can have a new perspective on the products—but lexicons are universalizable: they are "intersubjectivity engines" that allow structured communication about the subjective, perceptual qualities of foods, beverages, and other consumer products [5,12,13].

Because it requires human attention, the scale of DA is necessarily limited: the largest DA studies usually include no more than about 100 products. In comparison, there are thousands or even tens of thousands of whiskies currently on the market [14], hundreds of thousands of wines, and similar variety in other specialty food markets, like coffee, chocolate, beer, tea, and so on. All of these products are sold partly or entirely based on the value of their sensory attributes, and these attributes are described in various publications such as reviews and commentaries. While manual comment analysis requires fewer work hours than DA, it is still impractical for large datasets like the RateBeer or Yelp review corpora [15,16]. Computational methods are needed to fully leverage the power of existing sources of food-descriptive data.

1.2. Natural Language Processing and Machine Learning

The field of Natural Language Processing (NLP) has quickly matured in the last several years with the boom of deep learning techniques. A number of techniques have been developed in NLP for the identification of relevant terms in freeform text, but accomplishing this task in any given application is usually challenging due to the importance of domain-specific language and unique words not well-represented in language references like thesauruses.

Previous computational linguistic analysis of flavor in food descriptions has relied on having text with a high density of flavor descriptors [10,11], which is not always the case [8]. There are no published tools designed specifically for the identification of flavor descriptors. In this paper, we will prototype and test such a tool using a Long Short-Term Memory (LSTM) deep neural network [17] trained on manually-annotated whisky reviews. The architecture used is based off of a keyword extraction tool developed for identifying skills in resumes [18].

We believe that the flexibility of the Long Short-Term Memory (LSTM) deep neural network architecture in capturing context in natural language will allow the identification of unique language lexicons from whisky reviews. Free comments (reviews) are the most natural way for humans to describe their food experiences, but they are very hard to systematically analyze, especially in volume. The use of LSTMs presents an alternative to hand-coding and increases the volume of data we can meaningfully deal with. The ability to take advantage of previously written reviews to build lexicons with minimal human intervention has much value. Our investigation is a use-case scenario for LSTMs, but not the only one. This kind of architecture (and related ones) can probably solve many problems in Sensory Science (including sentiment and synonymy/similarity) as Sensory Science has seen very limited use of NLP.

1.3. Objectives

Our overarching aim is to be able to use preexisting reviews to create a flavor language by identifying and extracting the unique descriptors. Since the world of food and beverage descriptions is obviously a large and heterogeneous domain, we use whisk(e)y as a case study for this domain, as it is an important economic product [14] and one without an authoritative flavor language that comprehensively covers the many increasingly-relevant product styles [19]. This approach should, in principle, be applicable to any product with a large corpus of free-text description available, such as the flavor of other foods, textile feel, and perfume aroma.

To determine whether it is possible to separate flavor-descriptive terms from those with no sensory meaning in written descriptions of food experiences, such as those found on product review blogs, we created a data pipeline that uses NLP techniques to take freeform text whisky reviews and extract descriptors, which can be used to create a lexicon for each whisky: a list of characteristic descriptors or a descriptive representation. The results can be used to identify relationships between descriptors and allow us to begin understanding the flavor language of whiskies.

Therefore, our contributions are threefold:

- An interactive annotation tool to facilitate identifying descriptors and non-descriptors, which allows the creation of a training set for a deep learning architecture;
- The deep learning architecture for our problem domain; and
- The pipeline for data preparation, annotation, training, and testing not seen before in Sensory Science.

2. Materials and Methods

The training of a neural network requires a training set of positive and negative examples—in this case, words that are descriptive of flavors as opposed to other words in product reviews that do not describe flavor. To minimize the burden of manual annotation, we developed an interactive visual tagging tool. Full-text reviews were preprocessed to provide a list of potentially descriptive word forms for human annotation as descriptive or non-descriptive, and then individual instances of these descriptive and non-descriptive words in context were used to train and test the proposed LSTM architecture.

2.1. Data Collection

The dataset we used to train our model contains a total of 8036 full-text English whisky reviews scraped from four websites: WhiskyAdvocate (WA; 4288 reviews), WhiskyCast (WC; 2309 reviews), The Whiskey Jug (WJ; 1095 reviews), and Breaking Bourbon (BB; 344 reviews). WA and WC reviews are from websites affiliated with a magazine and podcast, respectively, and written by professional reviewers scoring whiskies from around the world and providing tasting notes. BB and WJ are smaller "semi-professional" review websites more focused on American whiskey. BB and WA have multiple named reviewers writing the tasting notes for their websites, while WC and WJ each have a single named reviewer.

WC, WJ, and BB were scraped using beautifulsoup4 in Python v3.7, while WA was scraped using the rvest package v0.3.2 in the R Statistical Environment v3.5.3. A combination of GET-specific scrapers were used to collect the review-containing URLs from the various sites, and then the content was collected with site-specific scrapers. When possible, page formatting was used to collect metadata about the products being reviewed such as country of production or the proportion of alcohol by volume (ABV), as well as the review itself, but the metadata were not consistent across sites.

2.2. Data Preparation

After collection, each review was converted from full text (excluding the title and metadata elements such as the date of publication) into a list of potentially-descriptive word base forms called "lemmas" that occurred in each review using the workflow described in [19]. Briefly, the reviews were first tokenized, or converted into an ordered list of individual words and punctuation. Each token was tagged with a part-of-speech (POS) label such as "adjective" or "punctuation", and all tokens other than nouns, adjectives, and a small whitelist of verbs were removed. The remaining tokens were lemmatized, or converted into their base form (e.g., "drying" to "dry").

This was done using Spacy v2.1.8 in Python v3.7. The pretrained model en_core_web_sm v2.1.0 was used as the basis for calculating predictions and R package cleanNLP v3.0 was used in R v3.5.3 to convert the data to a tabular CSV format.

These lemmas were used as the list of potential descriptors for manual annotation with our interactive visual tagging tool (described in the next section). The frequency of occurrence (as an adjective or noun) for each lemma was used to prioritize more common lemmas for annotation.

2.3. Interactive Tagging Tool

To create examples of descriptive and non-descriptive words in context for this study, human annotators used a browser-based interactive tagger tool based on a word cloud visualization, seen in Figure 1. The tool was built for this purpose in Javascript and HTML5 using jQuery v3.4.1. A CSV file of token frequencies is uploaded from the user's local storage, the most common terms are rendered into a word cloud using jQCloud v2.0.3, and the user assigns words to the descriptor (1) or non-descriptor (0) classes using one of three interaction modes. The central wordcloud display (Figure 1B) repopulates with progressively less common words as the user assigns words to classes, and the resulting corpus of labeled words is exported along with unlabeled words as a CSV file to the user's local storage. Up to 50 words can be displayed in the central panel at a time, based on the rendering algorithm described in [20]. Fancybox v3.5.7 is used to display tooltips.

With a low learning curve, the user is able to sift through the text in a timely manner and create a human-annotated list of positive and negative examples. The user is then able to save the corpus of labeled words to a comma separated value (CSV) file.

2.4. Gold Standard Annotations

We asked four annotators (A, B, C, and D) from Food Science to use the interactive tagger tool to create an annotated training set. The annotators were chosen based on their expertise in Sensory Science, a sub-field of Food Science. Annotators A and B were involved with annotating all the datasets, while C was a tiebreaker for datasets WA and WC and D was a tiebreaker for BB and WJ. A lemma was deemed a descriptor if it was tagged as such by two out of the three annotators; otherwise, the lemma was tagged as not being a descriptor. As such, the number of annotators was chosen so a best two out of three consensus could be achieved. This is important as it provides a more accurate set of labeled annotations and is a common practice in both corpus annotation in NLP [21] and in the analysis of freeform comments in sensory science survey research [22]. A total number of 1794 lemmas (499 descriptive, 1295 non-descriptive) were tagged and used to create a training and test set. There were a total of 2638 unique descriptive/non-descriptive tokens tagged based on these lemmas (e.g., the lemma "fruit" could appear in the text as "fruity" or "fruits", i.e., a lemma could result in multiple tagged tokens). All individual occurrences of the tokens in context were used for training.

Interactive Tagging Tool

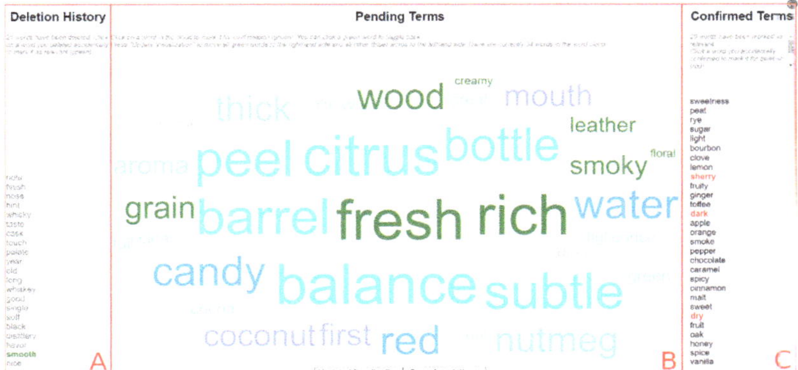

Figure 1. Interface for the Interactive Tagging Tool: (**A**) Non-descriptors (negative examples) are kept in the deletion history. (**B**) Then the most frequent words (up to 50) are shown in a word cloud format. (**C**) Confirmed descriptors (ones that are selected by the human operator) are stored in a confirmed terms list.

2.5. Word Embeddings

A word embedding is a representation of a word as a high-dimensional vector. The closer a pair of word vectors are in the high-dimensional space, the more the words are conceptually "similar" or "related". An input sequence for each word (i.e., potential descriptor) was created from the context and potential descriptor (unigram). Each potential descriptor and context word is assigned a 300 dimensional GloVe [23] word embedding. GloVe embeddings with 1.9 million tokens were used. A key note is that terms generally used in a domain specific language, such as those of whisky tasting notes, are not commonly used by the lay person, so this is a key consideration for domain-specific keyword extraction.

As illustrated in Figure 2, three words before and after were used as context, $n = 3$. If the context was less than three words; e.g., if the word was the first word of a sentence, then a PAD, a filler value, was used to signal no available context. The PAD value is assigned the zero vector. It is these input sequences that were used to train the model described in the next section.

2.6. Descriptor Extractor

We chose a uni-directional Long Short-Term Memory (LSTM) deep neural network architecture since it works well with the context of language. An LSTM is a Recurrent Neural Network (RNN) designed for modeling sequence data. LSTMs have a memory segment that can "remember" up to a certain degree of events in time. Hence, it works well with remembering context in language and the relationships between words. The context that a descriptor is found is essential to identifying what is or is not a descriptor. How a word is used can be a deciding factor.

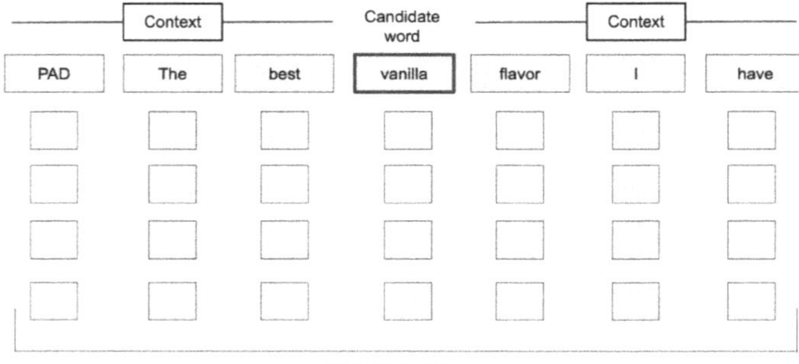

Figure 2. Conceptual example of an input sequence for training the model. The before and after context of a candidate word is extracted from a text review. This is then converted into a numerical representation using GloVe word embeddings for each context word and the candidate word with a PAD being the zero vector.

The architecture used was inspired by [18] and can be seen in Figure 3. There are two inputs, the context of a potential descriptor and the potential descriptor itself. Each is fed into a uni-directional LSTM of 256 units followed by a dense layer of size 128 units. These two dense layers are concatenated and fed forward to a series of decreasing dense layers ending with a binary softmax output layer that decides whether the input word is a descriptor or not. Defining the size of the context is flexible while we currently fix the descriptor (word) to a unigram as we observed most descriptors are single words. However, we chose a context of three words before and after a descriptor, n = 3.

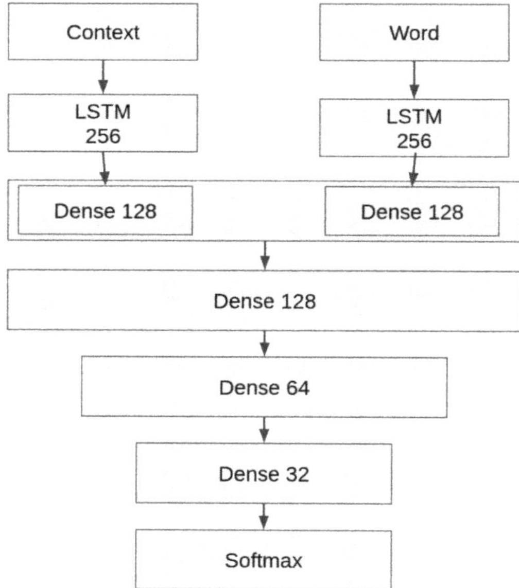

Figure 3. Deep Learning Architecture which is composed of two sets of input (context and a word) that feed into an uni-directional LSTM each. The rest of the architecture concatenates the LSTM layers and continues to merge dense layers with a softmax as the output.

We decided to use a traditional LSTM as a starting point for our keyword (descriptor) extractor. We wanted to see how well this model structure could perform before turning to more sophisticated model architectures in the future such as transformers [24]. As we will discuss in our results, the model architecture performed well.

The Descriptor Extractor was written in Python v3.6.9 with the deep learning architecture built using Keras v2.3.1 and Tensorflow v1.14.0. Comet.ml [25] was used to track different aspects of the model training, allowing us to provide detailed information presented in some of the figures in the Results section.

3. Results

Our experiments focused on testing our LSTM architecture as it was tailored to the problem space. For a comparison baseline, we chose parts-of-speech (POS) tagging since it closely reflects the characteristics of descriptors, which are generally adjectives and nouns. The POS approach is currently state-of-the-art for Sensory Science and therefore reflects a valid comparison [10,19].

Our first experiment combined the WA and WC datasets into one dataset for training and validation. We used an Adam optimizer with a learning rate of 0.0001, and our loss function was binary cross entropy with a batch size of 32. We had the BB and WJ datasets annotated in the same fashion as WA and WC so as to have a labeled test set. The results on the test set (accuracy/precision/recall/F1-score) can be seen in Table 1. The scores were lower compared to those of the training set. This made us rethink why this could be happening. We realized that an important difference between WA/WC and BB/WJ was that WA/WC were professionally written reviews, whereas BB/WJ were hobbyist reviews. There are likely different writing styles between the two kinds of reviews driving this difference in performance.

Table 1. Results from the test set (BB/WJ combined) when using WA/WC combined as the training and validation data.

	Accuracy (%)	Precision	Recall	F1-Score
Parts of Speech	47.863	0.209	0.946	0.3422
LSTM	90.000	0.779	0.422	0.547

We then combined all the datasets (WA, WC, BB, WJ) into one dataset and performed a train/test split of 80/20. We approached our methodology of splitting up the train and test sets differently. After combining WA, WC, BB, and WJ, we tokenized the reviews into words (tokens) and performed a train/test split on the tokens themselves instead of on a review basis. We also recorded the specific review it occurred in, the sentence within the review, and the position in the sentence. Therefore, instead of just performing a search for all locations of vanilla, for example, in all reviews for each training and test sets, we used the specifically tagged location for each instance of vanilla. From there, we were able to extract the context (n = 3 words before and after each token). Hence, we isolated where each instance of vanilla was for the respective training and test set. Combining the reviews to create a new train and test set allowed the model to be exposed to more variations in writing styles and hence become a more robust classifier.

Given the labeled descriptors/non-descriptors (2638 unique), we identified around 250K instances of the labeled words. As mentioned, we used a randomly chosen 80/20 split for training and testing. The total number of words used for training and validation were around 200K for training and around 50K for testing. For training, we removed punctuation but kept stop words as they are part of the context. Twenty percent of the training data were used for validation. Each training/test split contained a class ratio of 56% non-descriptors and 44% descriptors; hence, there was no class imbalance.

It was unclear as to how many epochs to train for. An epoch is the number of passes through the entire training dataset. We noticed that the accuracy converged to near 100%

quickly (within the first two epochs). To prevent overfitting, we ran the training for as many epochs as necessary until the loss did not improve and then plotted the training loss versus the validation loss to see how the training was behaving. We trained the model in increments of 5% use of the data up to using 100%. This resulted in 20 training sessions. This was done to observe how the model behaved given different numbers of training data. In practice, if a model performs poorly, the inclusion of more training data may improve results. We investigated the plots for 5%, 50%, and 100% (Figures 4–6, respectively). We observed that the loss values consistently crossed roughly around three epochs an then diverged (overfitting). This is marked as "Epoch 2" in the figures as Epoch 1 is really Epoch 0 in the figures. Hence, we chose to train for three epochs.

Training loss vs. validation loss for 5% of data

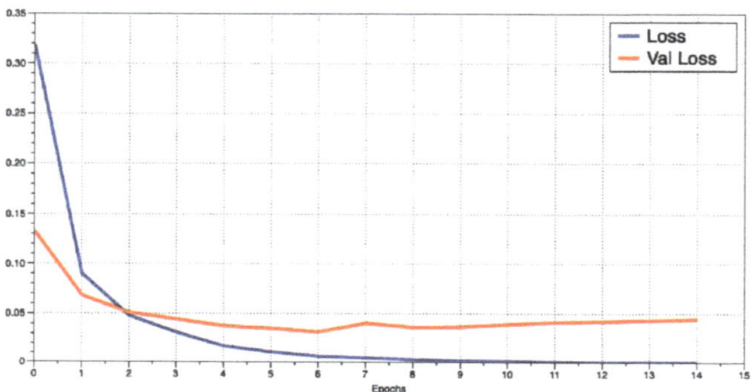

Figure 4. Training Loss versus validation loss using 5% of the training data where the x-axis is the epoch and the y-axis is the loss.

Training loss vs. validation loss for 50% of data

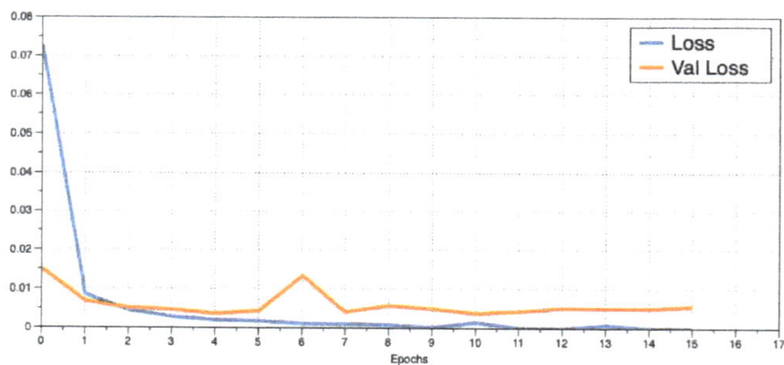

Figure 5. Training Loss versus validation loss for using 50% of the training data where the x-axis is the epoch and the y-axis is the loss.

After reviewing the loss and accuracy for each incremental iteration, we decided to report on using 100% of the training data. The gain from using all the data was small, e.g., loss difference of 0.00231 loss for 95% of the data versus 00.00238 for 100%. The difference in accuracy was equally minimal. Since the training with 100% of the data did not take long (around 5 min for three epochs using a desktop CPU), the small increase was still worth the extra training time. Figures 7 and 8 illustrate the training loss and validation loss, respectively, in which each loss (y-axis) is plotted in comparison to the percent of the data used in training (x-axis).

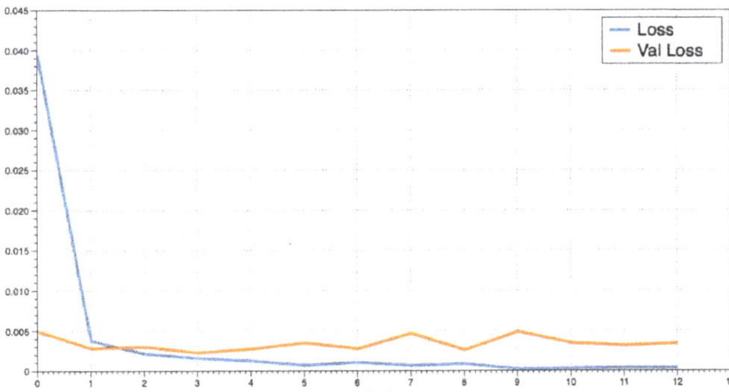

Figure 6. Training Loss versus validation loss for using 100% of the training data where the x-axis is the epoch and the y-axis is the loss.

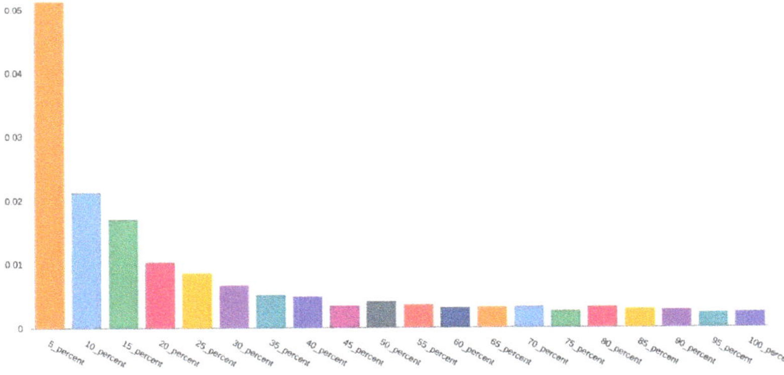

Figure 7. The loss for each percent of the training data used. x-axis is the specified percentage used and y-axis is the loss value.

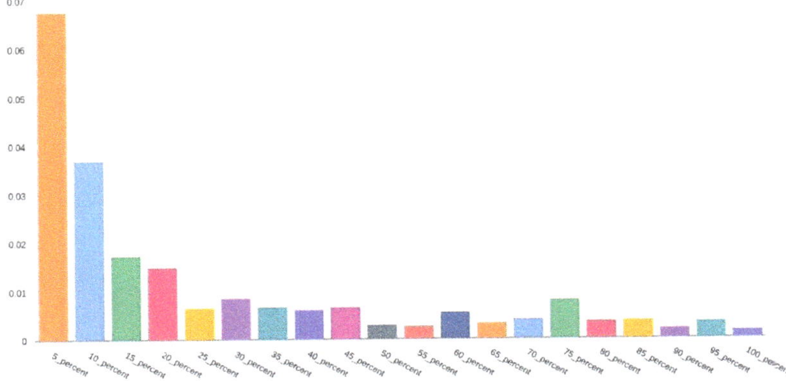

Figure 8. The validation loss for each percent of the training data used. x-axis is the specified percentage used and y-axis is the loss values.

We then use a parallel coordinate system (Figure 9) to illustrate how the use of more samples (higher percentage of data used) increases the batch accuracy, which corresponds to a lower loss and an overall higher accuracy. Note that the axis ranges for batch accuracy, loss, and accuracy are quite small.

The results from training can be seen in Table 2. Here the loss and accuracy are reported for the cases where available. Our model's training accuracy hovered at 99% with POS at 51.3%. The precision, recall, and F1-scores for the test set are presented in Table 3. One thing to note is that the recall is very high for POS. POS classifying is essentially saying "all nouns and adjectives are descriptors". In that case, there will be very few false negatives, because almost all descriptors ARE nouns and adjectives. Since recall = true positives/(true positives + true negatives), recall will be very high.

Parallel Coordinates of measurement relationships

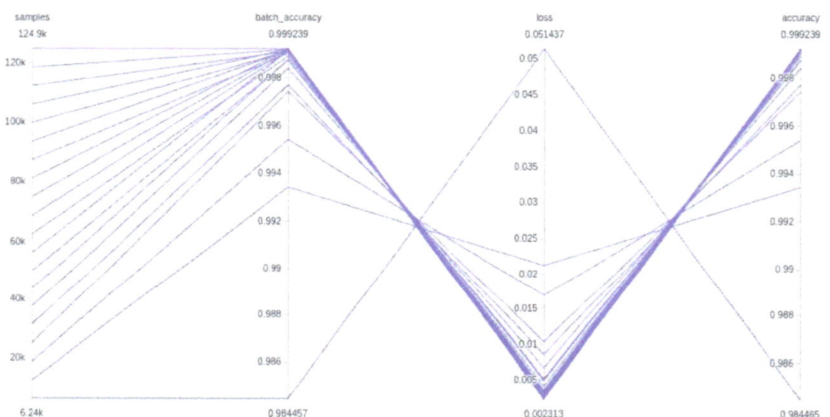

Figure 9. From left to right: The relationships between the number of samples used for training, the batch accuracy, the training loss, and the final accuracy.

Table 2. Training results from combining the datasets and splitting train/test along tokens.

100% of Data Used	Loss	Accuracy (%)
Train	0.00238	99.9234
Validation	0.00153	99.946
Test	N/A	99.910
Parts of Speech	N/A	51.344

Table 3. Test results from combining the datasets and splitting train/test along tokens.

	Precision	Recall	F1-Score
Parts of Speech	0.47410	0.92496	0.62688
LSTM	0.99883	0.99912	0.99898

An illustration of a case where the LSTM model struggles can be seen in Figure 10. The orange underlined words were identified by both a human annotator and the LSTM model. The blue ones were not identified by the LSTM model but were by the human annotator. The primary differences tend to be that the LSTM model only identifies the more descriptive word in bi-gram descriptor phrases (e.g., "banana chips") and will classify uncommon words, especially proper nouns, as descriptive, albeit with a low probability (e.g., "Redbreast", prediction of 71%). The challenge with bi-grams is a focus of future work discussed later, and the low probabilities can be addressed by using a filter threshold.

Figure 10 also demonstrates the difficulty in creating a tagged gold standard corpus, as words like "copper" and "heavy" that are not usually flavor words were annotated by the LSTM model as non-descriptors. In certain contexts, as in the idiosyncratic text of this review, these words are arguably capable of describing flavor. In the majority of reviews, however, "copper" instead describes the color (not flavor) of the spirit. The difficulty that these kinds of rarely descriptive words present for rapid annotation is also a focus of future work for the tagging scheme described in this paper.

Copper Caramel, dried apples, honey, copper, dried strawberries, buttery malt, spice and some banana chips. Banana pudding, caramel, Nilla Wafers, grassy malt, copper, vanilla taffy, spice and some citrus. Long → Dried fruit, spice, copper, malt and vanilla. Great balance, medium body and a light velvety feel. Coating and strangely heavy this delicious Irish whiskey sits on the senses with a far larger presence than you'd imagine for a chill filtered 40% ABV whiskey. The aroma is elegant and complex with a fruity and coppery essence. The palate slides through heavy and a tad oily with the same fruit and copper. The finish is a long fade of dried fruits, copper and some warm vanilla... it's good. The Jameson Pure Pot Still 15 Years was a distillery exclusive and precursor to the Redbreast 15, which itself started as a LE, and that same depth and elegance is noticed as is the copper that threads through it. Tasting it next to the RB 15, which is 46% and NCF, shows how this small change can have such a big impact. It's deeper and heavier and delivers an even more coating spirit. If you ever get the chance to taste them side-by-side do it, it's a fascinating comparison.

Figure 10. Prediction results for a random review. The orange underlined words were identified by the LSTM model. The blue ones were not identified.

We were also interested in viewing the relationships between words chosen as descriptors or non-descriptors. One approach to do this is to visualize the GloVe word embeddings using a t-SNE plot, an approach to visualize high-dimensional data in a two-dimensional space [26]. What results is a scatter plot visualization where distance between each word represents "similarity" based on word embeddings. The closer they are, the more conceptually similar.

We first plot the t-SNE for the annotated words within the training set. This can be seen in Figure 11 with a descriptor being a brown "X" and non-descriptor a blue dot. The words that were labeled in the training dataset create distinct clusters. This demonstrates that the human annotations created a well-defined cluster space, and hence, supports that the provided annotations were of good quality and the embeddings have enough understanding of flavor language to have captured it in the embedding space. The same can be said for the clusters for the test dataset (Figure 12). One may notice that there are some descriptor/non-descriptors "speckled" across the opposing cluster, i.e., words labeled as a descriptor are found within the non-descriptor cluster. This demonstrates that just because words are similar in an embedding space, their contextual meaning can vary. Zooming in to an example (Figure 13) of this, we see various terms that describe different aspects of bodies of water or climates. These words hold some form of similarities but are not deemed "equal" in a descriptive sensory sense.

In Figure 14, we see the embedding space for words that were predicted to be a descriptor (brown "X") or not a descriptor (blue dot). As can be seen, neatly defined clusters also emerge along with some "speckles". This provides support that the trained model is able to segment the space into descriptors and non-descriptors and hence carve out sensory terms. Another observation is that in all the t-SNE embedding plots, the non-descriptors outnumber the descriptors as was noted to be the case in the Related Works section by [8]. The actual descriptors are the minority, which makes sense as we observed that in the reviews, most words are non-sensory.

Training data word relationships

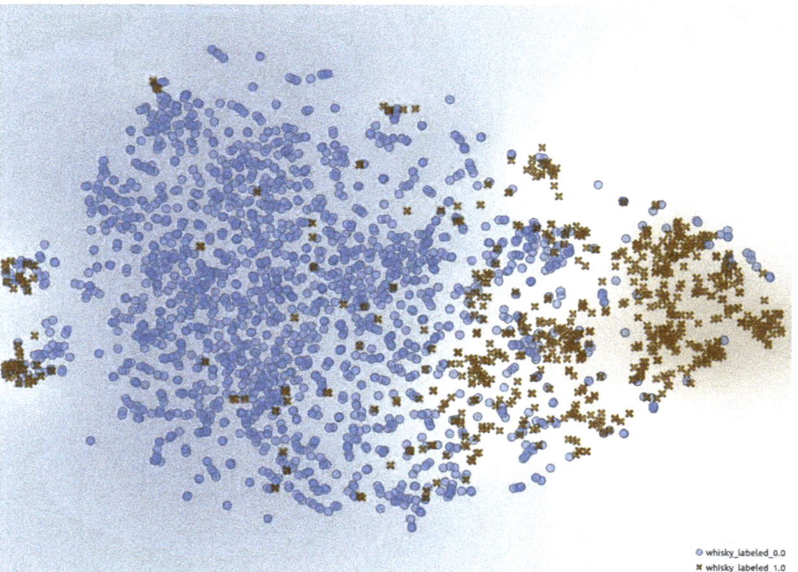

Figure 11. t-SNE representation of the training data where a blue dot represents a word labeled as a non-descriptor, and a brown "X" represents those that are descriptors.

Test data word relationships

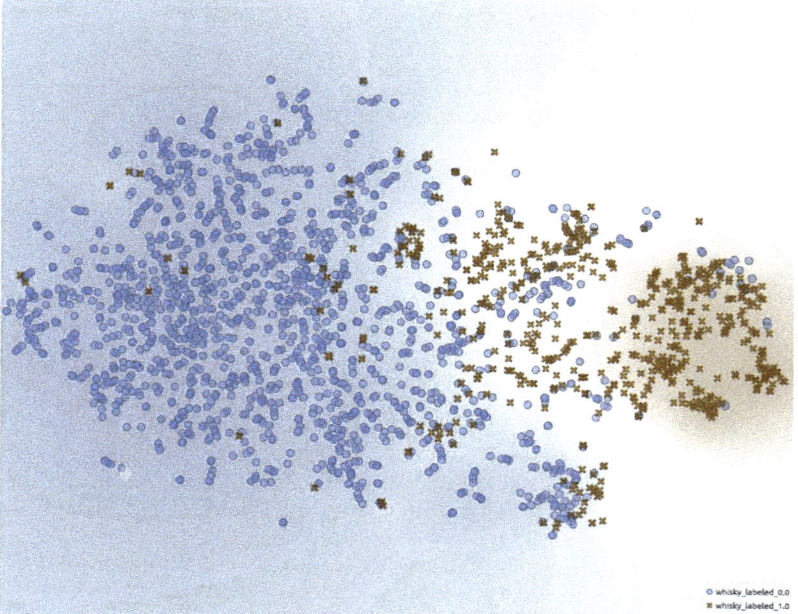

Figure 12. t-SNE representation of the test data where a blue dot represents a word labeled as a non-descriptor and a brown "X" represents those that are descriptors.

Closer look at word relationships

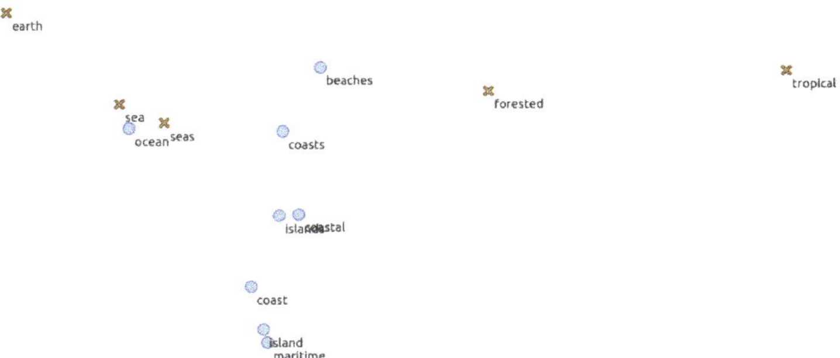

Figure 13. A zoom-in of a t-SNE plot to exemplify the separation of words into sensory and non-sensory despite their similarity of word embeddings. This shows that some words can hold a form of similarity but are not deemed "equal" in a descriptive sensory sense.

Predicted data word relationships

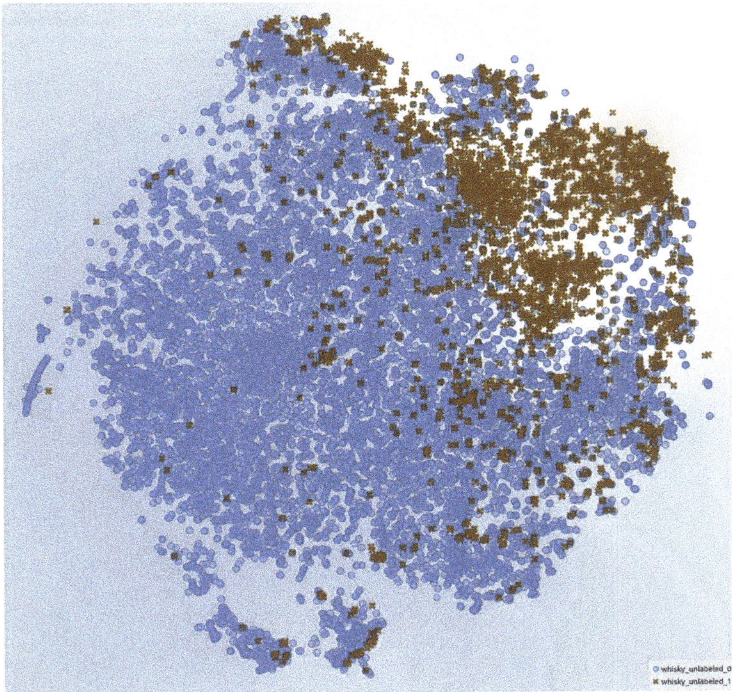

Figure 14. t-SNE representation of the words that were not annotated. A blue dot represents a word predicted as a non-descriptor, and a brown "X" represents those that are predicted as descriptors.

4. Discussion

We were able to build a deep learning model architecture based on LSTMs to provide descriptor identification within free-form text whisky reviews. Our results were very promising with training, validation, and test accuracies around 99%. The precision, recall, and F1-Scores were equally high. This is substantially higher than the current state of the art for Sensory Science. We were concerned about overfitting with such high scores, so we tracked the training and validation loss over many epochs, a common approach to detect overfitting. The tracking showed overfitting after three epochs, so we stopped our training at three epochs.

We were successfully able to automatically separate flavor-descriptive terms from those with no sensory meaning in written descriptions of food experiences (reviews). Our LSTM architecture was able to capture the language constructs that dictate what is and is not a descriptor. We view this as one of our key contributions.

Another key contribution is the pipeline for data preparation, annotation, training, and testing not seen before in sensory science. This opens the door for researchers in Sensory Science and Food Science in general.

We also introduced a novel interactive word tagging tool for creating a set of human-labeled descriptor/non-descriptor words. With multiple annotators using the tool, the set of human-labeled words provided an excellent training set. This supports the possibility of performing the same annotation with other datasets in other domains in order to facilitate the creation of a labeled set of words, hence training a model for those domains.

Visualizing the results using t-SNE revealed some interesting results. First, the embedding space of human labeled words by the interactive word tagger was segmented fairly cleanly into two clusters: those words that are descriptors and those that are not. This supports that the annotations are of high quality. Similarly, the embedding space for predicted words from the test set (non-annotated words) reveals two fairly clean clusters for descriptors and non-descriptors. This provides support that the trained model is able to learn the language structure of sensory terms.

These contributions result in some interesting and novel implications. We trained a model that can identify descriptors in texts, leading to the ability to create a lexicon for a whisky. Lexicons allow a comparison between whiskies: a $50 bottle of whisky can have a similar lexicon to that of a $300 bottle. This allows the consumer to "experience" the $300 whisky by trying the $50 whisky. Lexicons of whiskies can also map distinct descriptors to that of different metadata of the whiskies, such as age, region of origin, ingredients, and price. This can provide a foundation to perform predictive analyses, such as Random Forests. By fitting a model to predict continuous variables (e.g., bottle price, quality score) or classify products (e.g., region of origin) based on the presence or absence of flavor terms in the bodies of reviews, we can identify which flavors or flavor terms drive the price and consumer liking of whiskies or differentiate between product categories.

5. Limitations and Advantages

While our approach has had much success, there are a few limitations. The interactive tagger currently does not present the context of a word when the word is shown to the user. The context can influence whether some words are descriptive or not (e.g., "maple" in "a sweet maple aftertaste" vs. "this new offering from Maple Leaf Spirits"). Furthermore, unigrams are used to train the model with a context window of n = 3. Although this context window is based on the suggestion from [18], experimentation with other windows could be beneficial for our domain. The use of unigrams is also a limitation as some descriptors are not unigrams but phrases of two or more words. Studying beyond unigrams would be an important direction for future work.

The LSTM developed in this work is most immediately applicable to situations where researchers have some structured data about products (e.g., price, hedonic liking scores, chemical data, ingredient concentrations) and freeform product descriptions but no structured descriptive data. The presence or absence of the descriptors in each product, or the number of participants who used them, can be used very similarly to check-all-that-apply data in sensory analysis [10,27,28]. Freeform textual data is easier to collect than trained descriptive panel measurements, meaning that the use of this tool could reduce the barrier to entry for studies looking to relate production variables to resulting sensory properties or sensory properties to product liking and consumer behavior. The performance of the LSTM on descriptions of other foods (e.g., other spirits, wine, coffee, specialty meats and cheeses, casserole recipes) should be assessed before using the predicted descriptors for this kind of further analysis, but the interactive annotation tool developed in this work should reduce the amount of work necessary to tag small test sets and, if necessary, new training sets for these other domains.

6. Conclusions

In conclusion, we were successful in automatically detecting words as flavor-descriptive terms and separating them from non-sensory terms. Our developed deep learning architecture proved successful and opens the doors for further research into descriptive analysis.

For future work, we would like to perform testing of generalization between food domains, e.g., apply the model to cocktail and coffee descriptions. Currently, we use only unigram descriptors/non-descriptors. We would like to expand to using bi-grams and tri-grams as some sensory descriptors are phrases and not single terms (e.g., "wet dog", "red fruits"). Finally, we foresee the possibility of creating a word embedding dataset for the Food Sciences, i.e., an analogue to GloVe embeddings trained on sensory-specific descriptions.

Author Contributions: Conceptualization, C.M., L.H., and J.L.; methodology, C.M. and L.H.; software, C.M.; validation, C.M. and L.H.; formal analysis, C.M.; investigation, C.M.; resources, C.M., L.H., and J.L.; data curation, L.H.; writing—original draft preparation, C.M.; writing—review and editing, C.M., L.H., and J.L.; visualization, C.M. and L.H.; supervision, C.M.; project administration, C.M. and J.L.; funding acquisition, J.L. and C.M. All authors have read and agreed to the published version of the manuscript.

Funding: This research was funded by the Institute for Creativity, Arts, and Technology (ICAT) Major SEAD grant at Virginia Tech. The grant has no funding number; however, it does have a project page: https://icat.vt.edu/projects/2019-2020/major/seeing-flavors.html (accessed on 21 May 2021). The APC was funded by the Virginia Tech Libraries Open Access Subvention Fund and by author Lahne's overhead funds.

Data Availability Statement: The data presented in this study are available on request from the corresponding author. The data are not publicly available due to privacy concerns and copyright.

Acknowledgments: We would like to thank Jonathan Bradely of the Virginia Tech University Libraries for providing the prototype of the interactive tagging system. We would also like to thank our external annotators Brenna Littleson and Amanda Stewart.

Conflicts of Interest: The authors declare no conflict of interest. The funder had no role in the design of the study; in the collection, analyses, or interpretation of data; in the writing of the manuscript, or in the decision to publish the results.

Abbreviations

The following abbreviations are used in this manuscript:

DA	Descriptive Analysis
NLP	Natural Language Processing
LSTM	Long Short-Term Memory
WA	Whisky Advocate
WC	Whisky Cast
WJ	The Whiskey Jug
BB	Breaking Bourbon
CSV	Commab Separated Value
POS	Parts of speech

References

1. Buck, L.B. Olfactory Receptors and Odor Coding in Mammals. *Nutr. Rev.* **2004**, *62*, S184–S188. [CrossRef] [PubMed]
2. Heymann, H.; King, E.S.; Hopfer, H. Classical Descriptive Analysis. In *Novel Techniques in Sensory Characterization and Consumer Profiling*; Varela, P., Ares, G., Eds.; CRC Press: Boca Raton, FL, USA, 2014; pp. 9–40. Section 2.
3. Lawless, L.J.; Civille, G.V. Developing Lexicons: A Review. *J. Sens. Stud.* **2013**, *28*, 270–281. [CrossRef]

4. Drake, M.; Civille, G. Flavor Lexicons. *Compr. Rev. Food Sci. Food Saf.* **2003**, *2*, 33–40. [CrossRef] [PubMed]
5. Shapin, S. A taste of science: Making the subjective objective in the California wine world. *Soc. Stud. Sci.* **2016**, *46*, 436–460. [CrossRef] [PubMed]
6. Varela, P.; Ares, G. Sensory profiling, the blurred line between sensory and consumer science. A review of novel methods for product characterization. *Food Res. Int.* **2012**, *48*, 893–908. [CrossRef]
7. Ickes, C.M.; Lee, S.Y.; Cadwallader, K.R. Novel Creation of a Rum Flavor Lexicon Through the Use of Web-Based Material. *J. Food Sci.* **2017**, *82*, 1216–1223. [CrossRef]
8. Valente, C.C. Understanding South African Chenin Blanc Wine by Using Data Mining Techniques Applied to Published Sensory Data. Ph.D. Thesis, Stellenbosch University, Stellenbosch, South Africa, 2016.
9. Wishart, D. Classification of Single Malt Whiskies. In *Data Analysis, Classification, and Related Methods*; Kiers, H.A.L., Rasson, J.P., Groenen, P.J.F., Schader, M., Eds.; Series Title: Studies in Classification, Data Analysis, and Knowledge Organization; Springer: Berlin/Heidelberg, Germany, 2000; pp. 89–94. [CrossRef]
10. Bécue-Bertaut, M.; Pagès, J. Multiple factor analysis and clustering of a mixture of quantitative, categorical and frequency data. *Comput. Stat. Data Anal.* **2008**, *52*, 3255–3268. [CrossRef]
11. Moroz, D.; Pecchioli, B. Should You Invest in an Old Bottle of Whisky or in a Bottle of Old Whisky? A Hedonic Analysis of Vintage Single Malt Scotch Whisky Prices. *J. Wine Econ.* **2019**, *14*, 145–163. [CrossRef]
12. Hennion, A. Those Things That Hold Us Together: Taste and Sociology. *Cult. Sociol.* **2007**, *1*, 97–114. [CrossRef]
13. Shapin, S. The sciences of subjectivity. *Soc. Stud. Sci.* **2012**, *42*, 170–184. [CrossRef] [PubMed]
14. Lombardo, C. *Straight Up: Industry Revenue Will Steadily Grow as the Number of Independent Distillers Rises*; IBISWorld Industry Report OD4290; IBISWorld: Los Angeles, CA, USA, 2018.
15. McAuley, J.; Leskovec, J.; Jurafsky, D. Learning Attitudes and Attributes from Multi-aspect Reviews. In Proceedings of the 2012 IEEE 12th International Conference on Data Mining, Brussels, Belgium, 10–13 December 2012; pp. 1020–1025. [CrossRef]
16. Tao, D.; Yang, P.; Feng, H. Utilization of text mining as a big data analysis tool for food science and nutrition. *Compr. Rev. Food Sci. Food Saf.* **2020**, *19*, 875–894. [CrossRef] [PubMed]
17. Hochreiter, S.; Schmidhuber, J. Long short-term memory. *Neural Comput.* **1997**, *9*, 1735–1780. [CrossRef] [PubMed]
18. Ilin, I.; Chikin, V.; Solodskih, K. Deep Learning for Specific Information Extraction from Unstructured Texts. 2018. Available online: https://towardsdatascience.com/deep-learning-for-specific-information-extraction-from-unstructured-texts-12c5b9dceada (accessed on 20 May 2021).
19. Hamilton, L.M.; Lahne, J. Fast and automated sensory analysis: Using natural language processing for descriptive lexicon development. *Food Qual. Prefer.* **2020**, *83*, 103926. [CrossRef]
20. Ongaro, L.; White, D.; Sorel, D. jQCloud. Available online: https://mistic100.github.io/jQCloud/ (accessed on 29 October 2019).
21. Pustejovsky, J.; Stubbs, A. The Basics. In *Natural Language Annotation for Machine Learning*; O'Reilly: Sebastopol, CA, USA, 2012; Chapter 1.
22. Symoneaux, R.; Galmarini, M.V. Open-Ended Questions. In *Novel Techniques in Sensory Characterization and Consumer Profiling*; Varela, P., Ares, G., Eds.; CRC Press: Boca Raton, FL, USA, 2014; Chapter 12, pp. 307–332.
23. Pennington, J.; Socher, R.; Manning, C.D. GloVe: Global Vectors for Word Representation. In Proceedings of the 2014 Conference on Empirical Methods in Natural Language Processing (EMNLP), Doha, Qatar, 25–29 October 2014. [CrossRef]
24. Devlin, J.; Chang, M.; Lee, K.; Toutanova, K. BERT: Pre-training of Deep Bidirectional Transformers for Language Understanding. *arXiv* **2018**, arXiv:1810.04805.
25. Comet.ML Home Page. Available online: https://www.comet.ml/ (accessed on 21 May 2021).
26. van der Maaten, L.J.P.; Hinton, G.E. Visualizing High-Dimensional Data Using t-SNE. *J. Mach. Learn. Res.* **2008**, *9*, 2579–2605.
27. Meyners, M.; Castura, J.C. Check-All-That-Apply Questions. In *Novel Techniques in Sensory Characterization and Consumer Profiling*; Varela, P., Ares, G., Eds.; CRC Press: Boca Raton, FL, USA, 2014; Chapter 11, pp. 271–306.
28. Greenacre, M.J. *Correspondence Analysis in Practice*, 3rd ed.; CRC Press: Boca Raton, FL, USA, 2017.

Article

Insights into Drivers of Liking for Avocado Pulp (*Persea americana*): Integration of Descriptive Variables and Predictive Modeling

Luis Martín Marín-Obispo [1,†], Raúl Villarreal-Lara [1,2,†], Dariana Graciela Rodríguez-Sánchez [1], Armando Del Follo-Martínez [3], María de la Cruz Espíndola Barquera [4], Jesús Salvador Jaramillo-De la Garza [1], Rocío I. Díaz de la Garza [1] and Carmen Hernández-Brenes [1,*]

1 Tecnologico de Monterrey, Escuela de Ingenieria y Ciencias, Ave. Eugenio Garza Sada 2501, Monterrey, Nuevo Leon 64849, Mexico; A00815241@itesm.mx (L.M.M.-O.); raulvl@tec.mx (R.V.-L.); dariana@tec.mx (D.G.R.-S.); A00822822@itesm.mx (J.S.J.-D.l.G.); rociodiaz@tec.mx (R.I.D.d.l.G.)
2 SensoLab Solutions, Centro de Innovacion y Transferencia Tecnologica (CIT2), Ave. Eugenio Garza Sada 427, Monterrey, Nuevo Leon 64849, Mexico
3 ALFA, Centro de Tecnologia de Sigma Alimentos, Apodaca, Nuevo Leon 66629, Mexico; adelfollo@sigma-alimentos.com
4 Fundacion Salvador Sanchez Colin, CICTAMEX, S. C. Ignacio Zaragoza 6, Coatepec Harinas, Estado de Mexico 51700, Mexico; mespindolab@gmail.com
* Correspondence: chbrenes@tec.mx
† These authors contributed equally.

Citation: Marín-Obispo, L.M.; Villarreal-Lara, R.; Rodríguez-Sánchez, D.G.; Del Follo-Martínez, A.; Espíndola Barquera, M.d.l.C.; Jaramillo-De la Garza, J.S.; Díaz de la Garza, R.I.; Hernández-Brenes, C. Insights into Drivers of Liking for Avocado Pulp (*Persea americana*): Integration of Descriptive Variables and Predictive Modeling. *Foods* **2021**, *10*, 99. https://doi.org/10.3390/foods10010099

Received: 2 December 2020
Accepted: 29 December 2020
Published: 6 January 2021

Publisher's Note: MDPI stays neutral with regard to jurisdictional claims in published maps and institutional affiliations.

Abstract: Trends in new food products focus on low-carbohydrate ingredients rich in healthy fats, proteins, and micronutrients; thus, avocado has gained worldwide attention. This study aimed to use predictive modeling to identify the potential sensory drivers of liking for avocado pulp by evaluating acceptability scores and sensory descriptive profiles of two commercial and five non-commercial cultivars. Macronutrient composition, instrumental texture, and color were also characterized. Trained panelists performed a descriptive profile of nineteen sensory attributes. Affective data from frequent avocado adult consumers (n = 116) were collected for predictive modeling of an external preference map (R^2 = 0.98), which provided insight into sensory descriptors that drove preference for particular avocado pulps. The descriptive map explained 67.6% of the variance in sensory profiles. Most accepted pulps were from Hass and Colin V-33; the latter had sweet and green flavor notes. Descriptive flavor attributes related to liking were global impact, oily, and creamy. Sensory drivers of texture liking included creamy/oily, lipid residue, firmness, and cohesiveness. Instrumental stickiness was disliked and inversely correlated to dry-matter and lipids (r = −0.87 and −0.79, respectively). Color differences (ΔE_{ab}*) also contributed to dislike. Sensory-guided selection of avocado fruits and ingredients can develop products with high acceptability in breeding and industrialization strategies.

Keywords: avocado; cultivars; preference mapping; sensory evaluation; sensory descriptive analysis; consumer science

1. Introduction

Avocado fruits are now of high economic value, and thus, the food industry is showing a remarkable interest to enhance the production and processing of this crop [1,2]. According to the Statistical Division of the Food and Agriculture Organization of the United Nations (FAOSTAT), Mexico is the major producer and exporter of avocado worldwide. In 2018, Mexico's avocado production was 2,184,663 t, with a harvest area of 206,389 ha, representing a 2.17-billion-dollar market [3]. Nutritional characterization of avocado fruit identified many functional compounds, which include unsaturated fatty acids, vitamin E, tocopherols, ascorbic acid, B vitamins, carotenoids, potassium, phenols, antioxidants, phytosterols, acetogenins, and its derivatives containing a furan ring (called avocatins or avofurans), terpenoid glycosides, flavonoids, and coumarins [4–12]. The fruit's flesh is pale green to bright yellow in color; is smooth, buttery in consistency, and has an exquisite flavor and aroma [13]. Although the fruit is low in carbohydrates, it is high in lipids, proteins, and minerals [12,14]. Previous consumer studies identified relevant sensory attributes that characterized avocado pulp for texture (firmness, creaminess, buttery, smoothness, and watery) and flavor (grassy, bland, nutty, and buttery) [15,16]. However, avocado cultivars are reported to differ in their chemical profiles, which can influence their sensory characteristics and consequently, their acceptability. Although consumer liking of some avocado cultivars is reported [17,18], a trained panel-guided identification of the sensory attributes that impact a particular avocado cultivar's preference over another was not studied. The present work aimed to use predictive modeling to identify potential sensory drivers of liking for avocado pulp by evaluating the acceptability scores and sensory descriptive profiles of two commercial and five non-commercial cultivars. The study also characterized macronutrient composition, instrumental texture, and color as variables, to understand the liking and identify rapid assessment tools related to consumer acceptability.

2. Materials and Methods

2.1. Plant Material

Avocado (*P. americana*) cultivars, shown in Figure 1, were hand-picked (6 October 2017) from the Fundación Salvador Sanchez Colin—(CICTAMEX) experimental field, located in Coatepec Harinas, Estado de Mexico, Mexico (18° 55′ N, 99° 45′ W, 2240 m above sea level). Collected samples (10–12 kg of each cultivar) included five cultivars from the CICTAMEX collection and specimens from two commercially marketed cultivars (Hass and Fuerte). Fruits from all evaluated cultivars were harvested from the same orchard, from different locations within a single tree, selecting those at full physiological maturity (but unripe). Samples were air shipped in closed containers with activated charcoal. Upon arrival to the Centro de Biotecnologia-FEMSA (Tecnologico de Monterrey, Monterrey, NL, Mexico), all fruits were kept at 20 °C (85–90% relative humidity) for seven days, to complete the ripening process, until reaching an optimal state for consumption. Information related to horticultural race and general phenotype description of the cultivars used in the present study is summarized in Table 1.

Figure 1. Avocado (*Persea americana*) fruits including commercial and non-commercial cultivars.

Table 1. Morphological traits and genotype relationships of avocado cultivar samples at commercial ripeness.

Cultivar	Race [1,2]	Parentage/ Origin [2]	Fruit Main Phenotype Description	Fruit Weight (g) [3]	Pulp Yield (%) [3,4]	Fruit Length (cm) [3]
Ariete	M X G	Colin V-33/ Mexico	Ripe fruit is dark green, cream-colored pulp, and elliptical shaped seed [19].	441.8 (402.2–563.7) ^{a}	76.1 (75.8–76.1) ^{a}	15.0 ± 0.3 ^{a}
Colin V-33	M X G	Fuerte/ Mexico	Ripe fruit is dark green, cream-colored pulp, and triangular-shaped seed [19].	319.4 (274.1–400.9) ^{a}	73.9 (61.4–74.6) ^{b}	11.8 ± 0.5 ^{c}
Fuerte	M X G	Unknown/ Mexico	Ripe fruit is green, thick hull, cream-colored pulp, and triangular-shaped seed [20].	237.2 (201.4–274.4) ^{b}	73.6 (63.6–73.6) ^{b}	10.5 ± 0.4 ^{bc}
Fundacion II	M X G	Hass/ Mexico	Ripe fruit is dark purple color, cream-colored pulp, and circular shape seed [19].	190.1 (172.5–199.65) ^{c}	64.1 (57.1–64.1) ^{c}	7.4 ± 0.2 ^{d}
Hass	M X G	Unknown/ USA	Ripe fruit is dark purple, cream-colored pulp, with a smooth and creamy texture, and seed of small to medium circular shape [20].	198.0 (161.1–234.4) ^{b}	69.6 (68.2–69.9) ^{b}	9.8 ± 0.4 ^{c}
Jimenez II	M X G	Hass mutant/ Mexico	The fruit has a rough, leathery hull and not adhered to pulp, black color in the ripe stage [19].	195.0 (168.1–236.3) ^{b}	79.5 (60.1–79.5) ^{b}	10.7 ± 0.3 ^{bc}
Labor	M X G	Hass/ Mexico	Ripe fruit is dark green, cream-colored pulp with circular shaped seed [20].	336.0 (323.6–497.45) ^{a}	77.6 (69.4–77.7) ^{a}	13.6 ± 0.2 ^{ab}

[1] Race: M, (Mexican, *Persea americana var. drymifolia*), G (Guatemalan, *P. americana var. guatemalensis*).[2] Genotype assignation, parentage and origin reported by López-López, Barrientos-Priego, & Ben Ya'acov [21]; Rodríguez-López, Hernández-Brenes, & Díaz De La Garza [22]; Rendón-Anaya et al. [2]. [3] Values represent median (interquartile range) or mean ± SE for non-parametric or parametric data, respectively (n = 3–5). Different letters within the same column indicate significant differences, according to Kruskal-Wallis or LSD post-hoc test, for non-parametric or parametric data, respectively ($p < 0.05$); [4] g of pulp/100 g of total fruit's weight.

2.2. Physicochemical Analyses

The required number of avocado fruits (of each cultivar) were randomly selected to complete a sample composite of about 1.5 kg of pulp, based on their average weight and pulp content (Table 1). Avocado pulps were then manually pureed, and vacuum packaged in transparent nylon-polyethylene bags (Uline, Apodaca, NL, Mexico), containing approximately 250, 150, and 40 g of sample for their use in proximate composition, instrumental texture, and color determinations, respectively. The packaging material had a thickness of 5 μm, a standard barrier oxygen transmission rate of 63 cm^3/m^2 for 24 h at 23 °C, and 0% relative humidity, and a moisture vapor transmission rate of 4.8 g/m^2, 24 h at 37 °C at 90% relative humidity. Samples were stored at 4 °C, and physicochemical analyses were conducted within the following 24 h.

2.2.1. Proximate Macronutrient Composition

Moisture, protein, lipid, ash, sugar, and crude fiber content in avocado pulps were determined in triplicate, following the standard methods from the Association of Official Analytical Chemists International [23]. Total carbohydrate content was calculated by difference and dietary fiber was also determined in triplicate, using the AOAC methods 997.08 and 999.03.

2.2.2. Instrumental Texture Analyses

Instrumental texture determination of avocado puree samples was conducted using a TA-XTplus (Stable Micro Systems, Godalming, UK) texture analyzer. Measurements were performed using the TTC Spreadability Rig (HDP/SR) fixture, consisting of a set of male and female acrylic cones with 90° angles. Avocado puree packages were conditioned at

25 °C for 20 min before analysis. Samples were then filled into the female (lower) cone with a spatula, pressed lightly to eliminate air pockets (visible through the cone), and the surface was flattened. The female cone was fixed on the base holder of the texture analyzer. Protocol for cheese spread (Texture Exponent software version 6.1.11.0—Stable Micro Systems, Godalming, UK) was used in the determinations, with a 5 kg load cell (test speed—3.0 mm/s and post-test speed—10.0 mm/s). Distance traveled (23 mm) by the male cone was recorded, from its start point at 25 mm over the bottom of the female cone and until it was introduced into the sample, stopping when the final gap between the two cones was precisely 2 mm. Textural data were recorded as force in grams (g) versus time (s), and the software calculated the following instrumental parameters as output variables—firmness (g), work of shear (g s), stickiness (g), and work of adhesion (g s). Determinations were performed at controlled room temperature (25 °C) with five replicates per sample.

2.2.3. Instrumental Color Determinations

Instrumental color of avocado pulps from each cultivar were determined with a tristimulus Minolta CR-400 colorimeter (Konica Minolta Sensing Inc., Osaka, Japan), using a D75 illuminant at an observation angle of 10°. A standard white tile was used as a calibration reference. Readings for L^* (lightness), a^* (red-green axis), and b^* (yellow-blue axis) CIE*Lab* coordinates were recorded in five replicates ($n = 5$) for each cultivar. Variations of L^*, a^*, b^* (ΔL^*, Δa^*, Δb^*), and total color difference (ΔE_{ab}^*) were calculated for each cultivar using Hass commercial cultivar as a reference control. The following equation was used:

$$\Delta E_{ab}^* = \sqrt{(L_1^* - L_0^*)^2 + (a_1^* - a_0^*)^2 + (b_1^* - b_0^*)^2} \quad (1)$$

where L_0^*, a_0^*, and b_0^* included the reference values for control (Hass cultivar) and L_1^*, a_1^*, and b_1^* indicated values for cultivars. Values of $\Delta E_{ab}^* > 3.5$ units were considered as indicators that instrumental color differences were possibly perceived by an average observer [24].

2.3. *Sensory Analyses*

2.3.1. Sample Preparation

Fruits were weighed, washed, and soaked for 5 min in chlorinated water (200 ppm) for sanitization, and dried at room temperature for one hour. About 10 min before each sensory testing session (descriptive and affective tests), sample preparation was initiated; pulp was manually separated from peel and seed. Avocado fruits were randomly selected to obtain ~1 kg of pulp from each cultivar, based on their average weight and pulp content (Table 1). Pulps from each cultivar were hand-scooped and placed into plastic bags. Headspace was removed, and then the pulp was manually pureed within the bags, for 5 min, until color and texture were visually homogenous. For sensory evaluations, avocado puree samples (20 g) were placed in disposable soufflé cups (30 mL), identified with random three-digit numbers, and presented in random order to trained and untrained judges.

2.3.2. Sensory Descriptive Profiling

Ten trained panelists from SensoLab Solutions SC, a sensory and consumer science laboratory center, with over 500 h of descriptive experience in a wide variety of foods. conducted descriptive sensory profiling of nineteen sensory attributes using a 15-cm free scale. These attributes were obtained previously during two consensus sessions, where the panel as a group enlisted the most relevant attributes that characterized the studied samples. Five additional one-hour sessions were carried out to train the expert panel in the attributes that were obtained during the consensus. The ballot was designed using Fizz Forms (Biosystems, Couternon, France). All trials were conducted in individual sensory booths with white lighting and data were collected with FIZZ® Acquisition software version 2.50 (Biosystemes, Couternon, France). References and samples were rated using a 15-cm universal Spectrum™ line scale with 0 cm representing "none" and 15 cm representing "strong" [25]. Samples were presented with a random three-digit code, in random, monadic sequential order, and evaluated in triplicates, on four different days. Rinsing water and crackers were provided. No information about the test or samples was given to the panelists before or during the evaluations. Attribute definitions and references used in the evaluations are shown in Table 2.

Table 2. Sensory attributes, definitions, and references used in the descriptive analyses of avocado pulp.

Attribute	Definition	Reference [1] (Brand)
	AROMATIC FLAVORS	
Global impact	Maximum flavor intensity reached by the product.	Soybean oil (Nutrioli)
Lipidic complex	Flavor associated with any kind of fats.	Soybean oil (Nutrioli)
Creamy	Naturally occurring oil that binds flavors without tasting oily by a cream perception.	Mexican creole avocado puree with 5% heavy cream (Lala).
Oily	Flavor associated with oil.	Mayonnaise (Hellmann's)
Green/grassy	Aromatic characteristic of freshly cut leaves, grass, or green vegetables.	Freshly cut grass and Fresh lettuce
Fresh	Flavor associated with freshness.	Fresh lettuce
Seed	Character associated with chewing on seeds.	Avocado seed grinded
Earthy	A lingering earthy, musty flavor.	Sliced fresh mushroom
	BASIC TASTES	
Sweet	A fundamental taste of sucrose in water is typical.	2% sucrose solution
Sour	A fundamental taste of citric acid in water is typical.	0.05% citric acid solution
Bitter	A fundamental taste of caffeine in water is typical.	0.01% caffeine solution
	CHEMICAL FEELING FACTOR	
Astringent	Complex of drying, puckering, and shrinking sensations in the lower oral cavity.	Grape Juice (Welch's)
	TEXTURE	
Creamy/oily	Creamy or oily sensation in mouth.	Mayonnaise (Hellmann's)
Cohesiveness	Degree to which sample holds together in a mass.	Banana baby puree (Gerber)
Firmness	Degree of resistance to flow	Miracle Whip (Kraft-foods)
Fibers/strands	The degree to which fibers are present.	Mexican creole avocado puree
Spoon cover	Quantity of sample attached to the outer spoon surface when compressed against sample.	Heavy cream (Lala)
Spoon print	Print left by compressing a spoon in the sample.	Table cream (Nestle)
Lipid residue	Residual oily sensation after product is swallowed (oily residual).	Mayonnaise (Hellmann's)

[1] References were prepared approximately 24 h before a testing session, refrigerated overnight, and removed from the refrigerator 1 h before a testing session. Intensity based on a 15-point numerical scale, where 0 represents absence and 15 represents extremely strong. Avocado creole (unknown landrace) common in Northern Mexico was obtained from a local supermarket (Monterrey, NL, Mexico) and was used for calibration purposes, and reference intensities were established by the panel consensus.

2.3.3. Consumer Evaluations

Affective data were collected from $n = 116$ frequent avocado adult consumers (frequency > twice a week; 28% males; 72% females) within 18–51 years old (mean age 33.62 ± 12.73 years). Participants were previously recruited and were instructed to avoid eating or drinking anything but water at least two hours before the sensory evaluation. Consent forms provided participants with information on avocado samples. They were also asked for their willingness to participate in the study, as part of a graduate research project from the Department of Bioengineering, School of Engineering and Sciences of Tecnologico de Monterrey, Campus Monterrey, Mexico (Ethics ID: CSERDBT-0001). Participants evaluated appearance, texture, flavor, and overall liking, using a nine-level hedonic scale. The sessions were conducted at SensoLab Solutions SC, located at the Technology Transfer and Innovation Center of Tecnologico de Monterrey. Generally, 20–30 consumers participated in each evaluation session, and the study was conducted for three days. Participants received an economic incentive at the end of their participation.

2.4. Statistical Analysis

For physicochemical and instrumental determinations, normality of the data was evaluated by the Shapiro-Wilk test. The mean ± SE or median (interquartile range) was reported for parametric and non-parametric data, respectively. Determinations on the avocado composited samples included, proximate composition ($n = 2$), instrumental texture ($n = 5$), and instrumental color ($n = 5$), while the fruit's weight, length, and percent pulp contents were determined from individual specimens of each cultivar ($n = 3$–5). Analysis of variance (ANOVA) and mean separations were conducted with Fisher's least significant difference (LSD) post-hoc tests for parametric variables, and Kruskal-Wallis with Nemenyi post-hoc test for non-parametric variables. Significant differences were assessed at a $p < 0.05$. Pearson product-moment correlations for each pair of variables were also calculated to assess the relationships between sensory and physicochemical data. Statistical analyses were performed using the JMP software version 15.0 (SAS Institute, Cary, NC, USA).

For the descriptive data, ANOVA was performed using as a complete randomized design using product as a fixed effect and panelist as random effect, using Fizz Calculations software version 2.50 (Biosystems, Couternon, France). A post-hoc means comparison using Fischer's Protected LSD at a 95% confidence level was performed to determine significant differences [26,27]. For liking analysis, one-way ANOVA was performed and Fischer's Protected LSD as a post-hoc test at $p < 0.05$ level of significance using Fizz Calculations software version 2.50 (Biosystems, Couternon, France) [28].

External preference mapping methodology was used to relate the preferences shown by the consumers to descriptive sensory characteristics of the different avocado pulps. The first step consisted in mapping the pulps on the basis of their sensory descriptive characteristics. Principal Component Analysis (PCA) was used to construct a sensory descriptive biplot with all studied descriptive attributes (individuals run by principle means) by correlations (standardized). Components were retained if they explained at least 15% of the variance. The second step was the construction of the predictive models of external preference maps using consumer hedonic data, which were performed using the Fizz Calculations software version 2.50 (Biosystems, Couternon, France) and by the XLSTAT software (XLSTAT, 2020, Addinsoft, Germany). Consumer overall liking scores were regressed onto the product scores on the principal components of the sensory space included in the PCA biplot (obtained with the trained panel data), using a quadratic model. Quadratic surface model was selected, since it corresponded to the complete model, which allowed to take into account interactions between characteristics.

3. Results

3.1. Fruit Morphological Traits and Proximate Macronutrient Composition

Morphological traits of the sampled cultivars, as previously reported by CICTAMEX [19,20,29], were confirmed in the present work; their characteristics were documented in Table 1 and can be visualized in Figure 1. All cultivars used in the study, including Hass and Fuerte, were hybrids or selections of Mexican and Guatemalan races. The median fruit weight of non-commercial cultivar Jimenez II (195 g) was the closest in value to the weights of commercial cultivars Hass and Fuerte (198 and 237.2 g, respectively), followed by Fundacion II (190.1 g); while fruits from Colin V-33, Labor, and Ariete had average weights greater than 300 g.

Fruit length values for cultivars Colin V-33 and Jimenez II (11.8 and 10.7 cm, respectively) were not significantly different than those of commercial cultivars Hass and Fuerte (9.8 and 10.5 cm, respectively). While Ariete and Labor cultivars presented the fruits with the highest average lengths (13.6 and 15 cm, respectively). Fundacion II fruits presented the lowest length values (7.4 cm). Values of percent pulp yield are considered relevant parameters for commercial applications, since they represent the edible portion of the fruit. Pulp yields (Table 1) followed a similar trend than fruit lengths, thus the Colin V-33 and Jimenez II yield values (73.1 and 79.5%, respectively) were non-significantly different from those of commercial cultivars. While the longest cultivars Ariete and Labor had the highest median pulp yields (>76%), and Fundacion II (the shortest in length) had the lowest pulp yield (64.1%).

Proximate macronutrient composition, shown in Table 3, indicated that for five of the pulps, lipids were the primary macronutrient (>13%), closely followed by carbohydrates (>8%) and then proteins (1.3–2.2%). However, the carbohydrate contents for Jimenez II and Hass pulps were slightly higher or equal than the lipid contents, respectively. The moisture contents ranged between 54 and 73%, and the fiber and ash contents were less than 3%. Fuerte and Hass pulps contained significantly higher lipid contents (21.3 ± 0.2% and 17.7 ± 0.2%, respectively) and lower moisture levels (54.7 ± 0.2% and 57.7 ± 0.2%, respectively) than the other cultivars. Our results indicated that total lipid concentrations were inversely ($r = -0.88$, $p = 0.009$) related to moisture contents. Additionally, carbohydrate and sugar levels were both inversely and strongly correlated ($r = -0.95$, $p < 0.0012$) to moisture concentrations.

3.2. Descriptive Sensory Analyses of Avocado Pulps

Significant differences (LSD, $p < 0.05$) were observed for ten of the nineteen evaluated descriptors (Table 4). The sensory characteristics that differentiated the pulps the most were the attributes related to lipids' flavor impact and texture. While Hass, Jimenez II, Fuerte, and Colin V-33 were characterized for having the highest global flavor impact, Ariete had the lowest (LSD, $p < 0.05$). Creamy flavor was perceived in the highest impact for Colin V-33, Fuerte, Hass, and Jimenez II (LSD, $p < 0.05$). In texture, Hass had the strongest creamy/oily texture perception (LSD, $p < 0.05$). Additionally, Hass, Fuerte, and Jimenez II were also characterized for having the highest texture perception in firmness, cohesiveness, and lipidic residual, while Ariete, Fundacion II, and Labor had the lowest perception (LSD, $p < 0.05$). Additionally, the fiber strands attribute was different between samples (LSD, $p < 0.05$), being significantly higher in the Fundacion II cultivar.

A sensory PCA biplot was generated with all sensory descriptive attributes (Figure 2), where the first two dimensions explained 67.6% of variability in descriptive profiles of the tested avocado pulps. Sensory attributes were related to flavor, texture, and chemical factor sensations. Sensory attributes that loaded on Component 1 included flavor attributes such as lipid complex, creamy, global impact, oily, and earthy; it also included the texture descriptors lipidic residual, firmness, cohesiveness, creamy/oily, and spoon print. Component 2 involved flavor descriptors such as green/grassy, sweet, fresh, and sour.

3.3. Liking of Avocado Pulps by Consumers

The commercial preference of consumers towards the Hass cultivar was evidenced since it presented high liking scores that ranged between 7.1 and 7.2 hedonic points, as shown in Table 5. The most surprising results were obtained for the non-commercial cultivar Colin V-33 (6.9–7.2 hedonic points), since it ranked in the top liking group for all parameters and showed non-significant differences for appearance, texture, flavor, and overall liking when compared to Hass. The least overall liked cultivar was the non-commercial Fundacion II (5.6 points), and for the liking of appearance, commercial cultivar Fuerte (5.9 points) was also significantly lower (LSD, $p < 0.05$). Overall acceptability data for the rest of pulps ranged in the hedonic scale between 6.2 and 6.5 points; their values were slightly lower (but statically significant LSD, $p < 0.05$) than the most liked cultivars (Hass and Colin V-33).

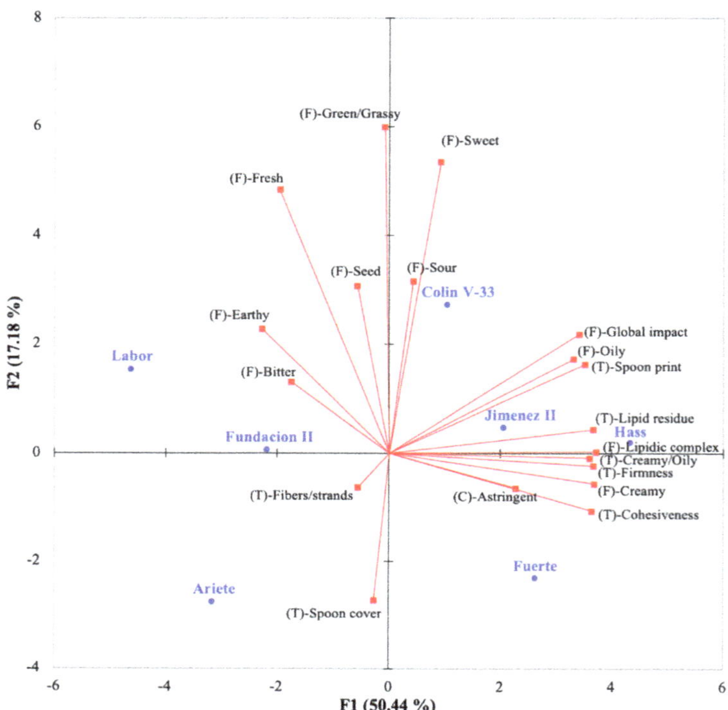

Figure 2. Principal component analysis (PCA) biplot of components F1 and F2, explaining 68% of the variance in the sensory descriptive profiles of seven avocado cultivars. Avocado samples are shown in blue (•), while vectors for sensory descriptive attributes are shown in red (■), and the descriptor names are shown in black. Sensory attributes were also classified with an abbreviation that indicated if they were related to flavor (F), sensory texture (T), or a sensory chemical sensation factor (C).

Table 3. Proximate macronutrient concentrations (g/100 g fresh weight (FW)) of seven avocado cultivars, including commercial and non-commercial samples.

Parameter	Ariete		Colin V-33		Fuerte		Fundacion II		Hass		Jimenez II		Labor	
Moisture	* 69.6	±0.0 ^{b}	65.9	±0.1 ^{c}	54.7	±0.2 ^{e}	66.6	±0.6 ^{c}	57.7	±0.2 ^{d}	58.4	±0.1 ^{d}	73.0	±0.0 ^{a}
Proteins	1.9	±0.0 ^{b}	1.6	±0.0 ^{c}	2.2	±0.1 ^{a}	1.9	±0.1 ^{b}	1.8	±0.0 ^{b}	1.5	±0.0 ^{c}	1.3	±0.0 ^{d}
Lipids	15.1	±0.1 ^{e}	16.0	±0.0 ^{d}	21.3	±0.2 ^{a}	13.8	±0.2 ^{f}	17.7	±0.1 ^{b}	17.0	±0.1 ^{c}	13.9	±0.0 ^{f}
Carbohydrates (CHOs)	8.4	±0.1 ^{c}	12.8	±0.1 ^{b}	16.6	±0.2 ^{a}	11.2	±1.8 ^{b}	17.8	±0.4 ^{a}	18.1	±0.1 ^{a}	8.6	±0.0 ^{c}
Sugars	2.8	±0.2 ^{d}	3.9	±0.0 ^{c}	5.0	±0.0 ^{b}	3.9	±0.0 ^{c}	5.3	±0.1 ^{a}	5.2	±0.1 ^{ab}	2.7	±0.1 ^{d}
Fiber dietary	2.9	±0.0 ^{a}	2.1	±0.0 ^{c}	2.1	±0.0 ^{bc}	1.9	±0.1 ^{d}	2.0	±0.0 ^{cd}	2.2	±0.0 ^{b}	1.7	±0.0 ^{e}
Ash	2.1	±0.0 ^{c}	1.7	±0.0 ^{d}	3.0	±0.1 ^{a}	2.7	±0.0 ^{b}	3.0	±0.0 ^{a}	2.8	±0.1 ^{b}	1.5	±0.0 ^{e}
Ratio Lipids/CHOs	1.8	±0.0 ^{a}	1.3	±0.0 ^{bc}	1.3	±0.1 ^{b}	1.3	±0.2 ^{bc}	1.0	±0.0 ^{cd}	0.9	±0.1 ^{d}	1.6	±0.0 ^{a}
Energy (kcal/100 g FW)	176.7	±0.1 ^{e}	201.3	±0.4 ^{d}	267.2	±0.3 ^{a}	176.5	±4.8 ^{e}	237.9	±0.5 ^{b}	231.2	±1.0 ^{c}	165.0	±0.2 ^{f}

* Values represent mean ± SE ($n = 2$). Different letters within the same row indicate that the means are significantly different, according to the LSD test ($p < 0.05$).

Table 4. Intensity ratings of nineteen descriptive attributes for seven commercial and non-commercial avocado cultivars.

Descriptor	Ariete	Colin V-33	Fuerte	Fundacion II	Hass	Jimenez II	Labor
(F) [1] Global impact	* 2.3 ±0.1 ^{d}	2.7 ±0.1 ^{ab}	2.6 ±0.1 ^{ab}	2.5 ±0.1 ^{bc}	2.8 ±0.1 ^{a}	2.7 ±0.1 ^{ab}	2.4 ±0.1 ^{cd}
(F) Lipidic complex	2.1 ±0.1 ^{cd}	2.4 ±0.1 ^{b}	2.5 ±0.1 ^{ab}	2.2 ±0.1 ^{c}	2.7 ±0.1 ^{a}	2.5 ±0.1 ^{ab}	1.9 ±0.1 ^{d}
(F) Creamy	1.8 ±0.1 ^{b}	2.2 ±0.1 ^{a}	2.4 ±0.1 ^{a}	1.8 ±0.1 ^{b}	2.4 ±0.1 ^{a}	2.4 ±0.1 ^{a}	1.6 ±0.1 ^{b}
(F) Oily	1.8 ±0.1 ^{cd}	2.3 ±0.1 ^{ab}	2.0 ±0.1 ^{bc}	2.0 ±0.1 ^{bc}	2.3 ±0.1 ^{a}	2.1 ±0.1 ^{ab}	1.6 ±0.1 ^{d}
(F) Green/Grassy	2.0 ±0.1 ^{c}	2.3 ±0.1 ^{ab}	2.1 ±0.1 ^{bc}	2.2 ±0.1 ^{abc}	2.2 ±0.1 ^{abc}	2.3 ±0.1 ^{ab}	2.3 ±0.1 ^{a}
(F) Fresh	1.4 ±0.1 ^{b}	1.6 ±0.1 ^{ab}	1.4 ±0.1 ^{b}	1.6 ±0.1 ^{ab}	1.5 ±0.1 ^{b}	1.6 ±0.1 ^{ab}	1.8 ±0.1 ^{a}
(F) Seed	1.6 ±0.1 ^{ab}	1.7 ±0.1 ^{a}	1.3 ±0.1 ^{b}	1.5 ±0.1 ^{ab}	1.5 ±0.1 ^{ab}	1.7 ±0.1 ^{a}	1.6 ±0.1 ^{ab}
(F) Earthy	1.0 ±0.1 ^{a}	1.0 ±0.1 ^{a}	0.8 ±0.1 ^{a}	1.0 ±0.1 ^{a}	0.8 ±0.1 ^{a}	0.9 ±0.1 ^{a}	0.9 ±0.1 ^{a}
(F) Sweet	0.6 ±0.1 ^{b}	1.0 ±0.1 ^{a}	0.7 ±0.1 ^{b}	0.7 ±0.1 ^{b}	0.8 ±0.1 ^{ab}	0.7 ±0.1 ^{b}	0.8 ±0.2 ^{ab}
(F) Sour	0.1 ±0.0 ^{b}	0.2 ±0.0 ^{a}	0.1 ±0.0 ^{ab}	0.1 ±0.0 ^{ab}	0.1 ±0.0 ^{ab}	0.1 ±0.0 ^{ab}	0.1 ±0.0 ^{ab}
(F) Bitter	0.2 ±0.0 ^{a}	0.2 ±0.0 ^{a}	0.2 ±0.0 ^{a}	0.2 ±0.0 ^{a}	0.2 ±0.1 ^{a}	0.2 ±0.0 ^{a}	0.2 ±0.0 ^{a}
(C) Astringent	1.9 ±0.1 ^{ab}	2.0 ±0.1 ^{ab}	2.2 ±0.1 ^{a}	1.9 ±0.1 ^{ab}	1.9 ±0.2 ^{a}	2.0 ±0.1 ^{ab}	1.8 ±0.1 ^{b}
(T) Creamy/Oily	3.2 ±0.2 ^{cd}	3.7 ±0.2 ^{bc}	3.9 ±0.3 ^{b}	3.2 ±0.2 ^{cd}	4.6 ±0.2 ^{a}	3.8 ±0.3 ^{b}	2.9 ±0.2 ^{d}
(T) Cohesiveness	4.8 ±0.3 ^{c}	5.6 ±0.3 ^{b}	6.8 ±0.3 ^{a}	5.0 ±0.2 ^{bc}	7.0 ±0.2 ^{a}	6.4 ±0.2 ^{a}	4.4 ±0.2 ^{c}
(T) Firmness	4.7 ±0.3 ^{cd}	5.4 ±0.3 ^{bc}	6.0 ±0.3 ^{ab}	5.0 ±0.2 ^{cd}	6.5 ±0.3 ^{a}	6.1 ±0.3 ^{ab}	4.6 ±0.2 ^{d}
(T) Fibers/strands	3.5 ±0.4 ^{bc}	3.3 ±0.4 ^{c}	4.2 ±0.4 ^{b}	6.7 ±0.3 ^{a}	4.2 ±0.4 ^{b}	3.4 ±0.4 ^{c}	3.7 ±0.4 ^{bc}
(T) Spoon cover	6.6 ±0.5 ^{a}	6.2 ±0.5 ^{ab}	6.6 ±0.5 ^{a}	5.6 ±0.4 ^{ab}	5.1 ±0.5 ^{b}	6.4 ±0.5 ^{a}	5.8 ±0.4 ^{ab}
(T) Spoon print	10.3 ±0.5 ^{c}	11.4 ±0.5 ^{abc}	11.0 ±0.7 ^{abc}	10.5 ±0.5 ^{bc}	11.8 ±0.6 ^{a}	11.7 ±0.4 ^{ab}	10.1 ±0.5 ^{c}
(T) Lipid residue	3.1 ±0.3 ^{c}	3.9 ±0.3 ^{b}	4.1 ±0.3 ^{ab}	3.1 ±0.2 ^{c}	4.6 ±0.2 ^{a}	4.0 ±0.2 ^{ab}	3.1 ±0.2 ^{c}

[1] Letters in parenthesis indicate attribute type designated as flavor (F), texture (T), and chemical factor sensation (C). * Values represent mean ± SE (10 trained panelists by triplicate, $n = 30$). Different letters within the same row indicate that the means are significantly different, according to LSD test ($p < 0.05$).

Table 5. Consumer acceptability scores for overall liking, flavor, appearance, and texture of seven avocado cultivars; including commercial and non-commercial samples.

Descriptor	Ariete	Colin V-33	Fuerte	Fundacion II	Hass	Jimenez II	Labor
Appearance liking	* 6.2 ±0.2 ^{cd}	7.2 ±0.1 ^{a}	5.9 ±0.2 ^{d}	5.8 ±0.2 ^{d}	7.1 ±0.1 ^{ab}	6.6 ±0.2 ^{c}	6.6 ±0.2 ^{bc}
Texture liking	6.4 ±0.2 ^{b}	7.2 ±0.1 ^{a}	6.4 ±0.2 ^{b}	5.5 ±0.2 ^{c}	7.1 ±0.1 ^{a}	6.3 ±0.2 ^{b}	6.5 ±0.1 ^{b}
Flavor liking	6.2 ±0.2 ^{cd}	6.9 ±0.2 ^{ab}	6.4 ±0.2 ^{cd}	5.8 ±0.2 ^{e}	7.3 ±0.1 ^{a}	6.5 ±0.2 ^{bc}	6.0 ±0.2 ^{de}
Overall liking	6.3 ±0.2 ^{b}	7.1 ±0.1 ^{a}	6.4 ±0.2 ^{b}	5.6 ±0.2 ^{c}	7.2 ±0.1 ^{a}	6.5 ±0.2 ^{b}	6.2 ±0.2 ^{b}

* Values represent mean ± SE ($n = 116$). Different letters within the same row indicate that means are significantly different, according to the LSD test ($p < 0.05$).

3.4. Preference Mapping of Consumer Acceptability and Descriptive Sensory Attributes

As previously mentioned, significant differences were observed in the liking of consumers for the seven studied cultivars. Overall acceptability scores for the seven pulps ranged from 5.6 to 7.2 in a nine-point hedonic scale, and generated three distinctive groups (LSD, $p < 0.05$). The external preference map shown in Figure 3 was obtained by modeling the overall acceptability scores over the descriptive map, where the best fit was obtained using a quadratic model ($R^2 = 0.98$). Furthermore, the overall liking scores were highly correlated to appearance, texture, and flavor variables ($r = 0.87$, 0.96, and 0.96, respectively, $p < 0.05$).

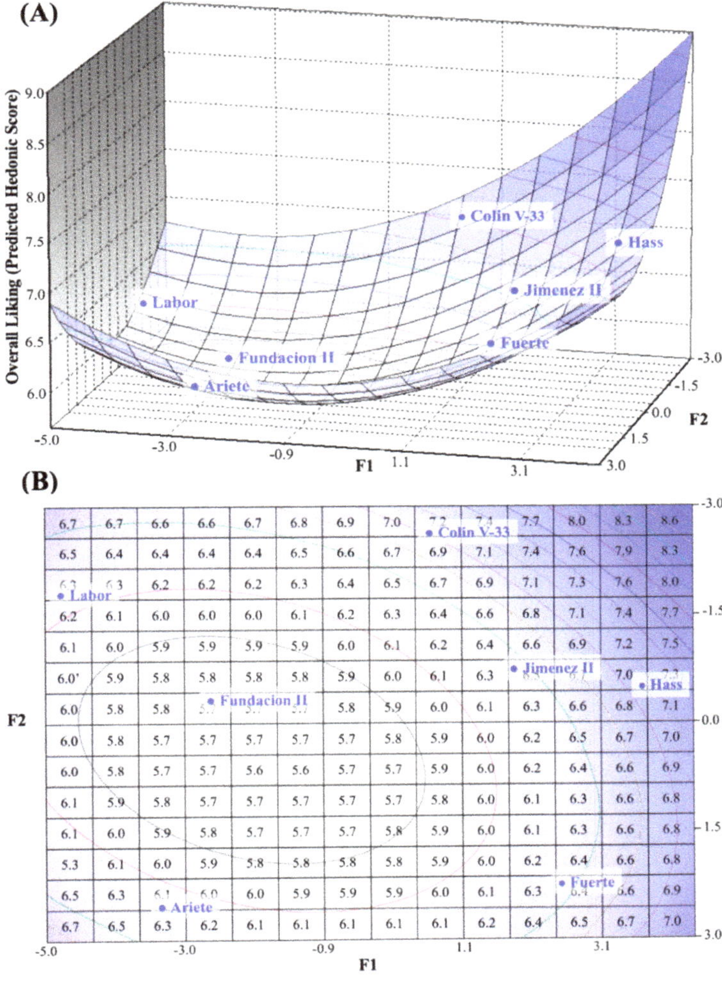

Figure 3. Tridimensional (**A**) and bidimensional (**B**) external preference map obtained by quadratic modeling of the overall liking of frequent avocado consumers (n = 116) for seven avocado cultivars, and placement of the affective data within sensory descriptive space. Principal component biplot of sensory descriptive data (components F1 and F2) used in construction of external preference map is shown in Figure 2. * Scores in the bidimensional map (B), represent the predictive hedonic values.

External preference mapping of affective data (Figure 3) was aligned with the descriptive map (Figure 2) to gain further understanding on the sensory attributes that drove the preference of the evaluated avocado pulps. According to the results, the sensory attributes that appeared responsible for the driving of like (areas shown in circles in Figure 3A,B) included the flavor descriptors of global impact, oily, and the texture attributes of creamy/oily and firmness. The alignment of descriptive and affective data in the preference map (Figure 3 and Figure S1) also provided valuable insight into which pulps were most desirable for percentages of consumers. Hass, Colin V-33, and Jimenez II samples fell in a region of the map that was characterized by high-acceptability (90–100% of satisfied consumers), followed by a region of slightly lower but still high-acceptability rates (60–80% of satisfied consumers), in which cultivar Fuerte was located. None of the avocado pulps were located in the mid-acceptability region (20–60% satisfaction) but Fundacion II, Labor, and Ariete were located in the least-liked region in which consumer satisfaction percentages ranged from 0–20%.

3.5. Proximate Macronutrient, Instrumental Texture, Instrumental Color, and Sensory Relationships

The relationships between sensory affective and descriptive data, shown in Figure 3 and described in the previous section, were key to gain insight on the potential sensory drivers of liking. A second PCA was also constructed to visualize how the differences in proximate macronutrient composition and instrumental texture were related with the sensory descriptive attributes of the avocado pulps (Figure 4). As previously discussed, cultivars that were located in high-acceptability regions (Hass, Colin V-33, and Jimenez II) were associated with sensory attributes related to flavor, particularly flavors associated with lipid notes. Total carbohydrates and sugars were variables that appeared to be relevant to the differentiation of Hass, Colin V-33, and Jimenez II from other pulps, as they loaded in the same PCA quadrant (Figure 4).

Table 3 shows the lipids to carbohydrate ratios for the pulps; this parameter indicated that when the lipid content gets higher in relation to their carbohydrate content (as for the Ariete and Labor pulps), the balance in flavor sensory attributes in the PCA quadrant seemed to move away from the desirable intensities (Figure 3). However, as shown in Figure 4, the relationship between macronutrient composition and desirable sensory descriptive profiles was not simple. Cultivar Colin V-33, which was among the most liked by consumers, had similar lipid to carbohydrate ratios than Fuerte cultivar and the least liked Fundacion II cultivar (Table 3), but Colin V-33′s sensory sweetness scores were significantly higher (Table 4).

PCA biplot shown in Figure 4 also aided in the visualization of chemical components (Table 3), sensory attributes (Table 4), and instrumental texture parameters (Table 6) that differentiated avocado pulps. Most sensory texture attributes related to lipidic sensations in the mouth, assessed by trained panelists, loaded in the quadrant with the most desirable pulps (Hass, Jimenez II, and Colin V-33). Relevant sensory texture attributes included lipidic residual, firmness, creamy/oily, and spoon print. Some instrumental texture parameters were noted to be correlated with some sensory texture descriptive attributes such as cohesiveness, which inversely correlated with instrumental stickiness ($r = -0.75$). Additionally, instrumental stickiness showed a significant ($p = 0.01$) and direct correlation with moisture contents ($r = 0.88$). As shown in Figure 4, both the stickiness and moisture vectors were characteristics associated with the least liked cultivar Labor.

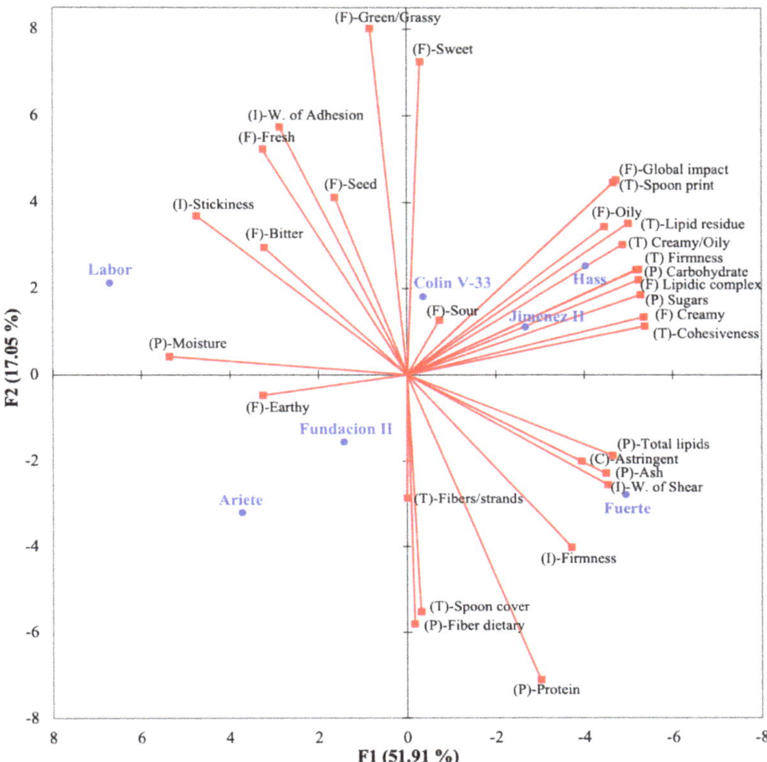

Figure 4. Principal component analysis (PCA) biplot of components F1 and F2, explaining 69% of the variance in the sensory descriptive profiles, proximate macronutrient composition, and instrumental texture analysis of the seven avocado cultivars. Avocado samples are shown in blue (•), while vectors for sensory descriptive attributes are shown in red (■), and the descriptor names are shown in black. Variables were also classified with an abbreviation indicating if they were related to sensory flavor (F), sensory texture (T), sensory chemical sensation factor (C), proximate macronutrient composition (P), or instrumental texture analysis (I).

Table 6. Instrumental texture parameters of seven commercial and no commercial avocado cultivars.

Cultivar	Firmness (g)	Work of Shear (g s)		Stickiness (g)		Work of Adhesion (g s)	
Ariete	* 433.6 (411.7–444.1) ^{b}	* 414.4	±6.4 ^{d}	−418.1	±3.2 ^{b}	−125.6	±2.7 ^{c}
Colin V-33	543.7 (499.6–590.4) ^{a}	546.1	±22.7 ^{b}	−480.5	±11.8 ^{c}	−147.3	±4.1 ^{c}
Fuerte	594.9 (570.3–618.6) ^{a}	626.9	±18.1 ^{a}	−653.6	±6.3 ^{f}	−170.5	±4.1 ^{d}
Fundacion II	568.8 (503.3–604.6) ^{a}	560.0	±24.2 ^{b}	−512.9	±5.6 ^{d}	−155.2	±6.2 ^{c}
Hass	434.5 (418.0–442.3) ^{b}	486.2	±12.4 ^{c}	−480.5	±7.6 ^{c}	−112.9	±2.7 ^{b}
Jimenez II	507.8 (498.5–532.6) ^{a}	574.2	±16.1 ^{b}	−55.3	±7.9 ^{e}	−127.3	±6.9 ^{b}
Labor	315.2 (307.2–339.1) ^{c}	331.6	±11.1 ^{e}	−323.7	±6.5 ^{a}	−94.3	±4.9 ^{a}

* Values represent median (interquartile range) and mean ± SE for nonparametric and parametric data, respectively ($n = 5$). Different letters within the same column indicate significant difference, according to the Kruskal-Wallis or LSD post-hoc test, respectively ($p < 0.05$).

The liking of commercial Fuerte cultivar was difficult to understand since it ranked in the second-best group for overall liking (Table 5 and Figure S1); however, its chemical and texture characteristics were different to those of Hass, Jimenez II, and Colin V-33.

In the PCA biplot space (Figure 4), Fuerte cultivar was associated with the vectors for instrumental firmness, total lipids, and work of shear; the latter was inversely related to moisture ($r = -0.77$).

Data on instrumental colorimetric parameters of avocado pulps were expressed as ΔL^*, Δa^*, Δb^* in relation to Hass cultivar (as reference control). Instrumental color differences among avocado cultivars are shown in Figure 5 and Table S2. Color variation values (ΔE_{ab}^*), also shown in Figure 5, were also calculated as quantitative parameters that integrated the ΔL^*, Δa^*, Δb^* values. Results indicated that the least liked pulps, Labor and Ariete, presented higher ΔL^* values ($\Delta L^* = +8.15$ and $\Delta L^* = +7.20$, respectively) indicating higher lightness values than Hass cultivar. Colin V-33, Fuerte, Fundacion II, and Jimenez II only showed minor variations in ΔL^*, denoting similarity to Hass cultivar. In contrast, Δa^* and Δb^* values, in reference to the Hass cultivar, were similar for Fuerte and the non-commercial cultivars, suggesting that the green and yellow chromaticity was similar among all pulps. Color differences (ΔE_{ab}^*) were the instrumental parameters that differentiated samples the most from Hass and the values ranged from 1.20 to 8.31 (Figure 5).

4. Discussion

4.1. Fruit Morphological Traits and Proximate Macronutrient Composition

Cultivars developed by the CICTAMEX foundation (Ariete, Colin V-33, Fundacion II, Jimenez II, Labor) were characterized morphologically, chemically, and compared to commercial cultivars (Hass and Fuerte) (Tables 1 and 3). In agreement with our findings, Cajuste-Bontemps et al. [29] and Alemán-Reyes et al. [19] reported that fruits from Colín V-33 cultivar presented similar pulp yields than those form Hass and Fuerte cultivars (Table 1). However, Colin V-33 had significantly higher fruit weights; similarly, prior authors reported high fruit weights (>319 g) for the same cultivar and described it as an unfavorable commercial characteristic since the calibers of greater demand oscillate between 200 and 300 g [19,29].

Results from proximate macronutrient analyses (Table 3) indicated that all sampled cultivars (Guatemalan X Mexican hybrids) were above the California Avocado Industry standard for minimum dry matter percentages of 20.8% set for Hass [30]. According to Yahia & Woolf [31], the 20.8% dry matter standard approximates a minimum oil content of 8%. Total lipid concentrations shown in Table 3 were found to be inversely correlated ($r = -0.88$) to moisture contents. Similarly, previous research reported that during avocado fruit development, the moisture levels declined, detailed as parallel increases in dry matter and lipid contents [32]. Other researchers that focused on the chemical characterization of avocado pulps observed that high moisture and low moisture in dry fruits contained lower lipid levels, but also showed lower levels of other macronutrients such as carbohydrates, sugars, and proteins [1,30].

4.2. Descriptive Sensory Analyses of Avocado Pulps

In the present study, sensory descriptive analyses showed that the attributes that differentiated the pulps the most were related to the lipids' impact on flavor and texture descriptors (Figure 2). In agreement, a positive correlation between oil content and palatability (flavor), as a unique sensory attribute, was previously reported for different commercial cultivars as determined by "super-critical tasters" that were very familiar with the avocado fruit and expected more of it than would the average consumer [33]. Other published sensory studies also performed the scaling of various sensory attributes in avocado samples with different chemical compositions, although not with trained panels. In a study conducted by Obenland et al. [15], avocado sensory attributes were defined by a consumer panel (n = 15–20), which also conducted affective testing of the same twelve avocado samples; data were used for the selection of eight main sensory attributes that were associated with the samples. All avocado samples included in their study were from the Hass cultivar, grown in different locations, and harvested on different years. The list of potential avocado descriptors was based on a previous study also conducted with the cultivar Hass [31]. Although foundational work for the selection of the eight main attributes present in avocado, with consumer evaluations, generated valuable knowledge [15]; the attributes were determined using only Hass cultivar, which might have possibly limited the sensory description of other attributes not present or pronounced in that particular cultivar. The main avocado sensory attributes identified in their work included four texture attributes (firm, creamy, buttery, smooth, and watery) and four flavor attributes (grassy, bland, nutty, and buttery). Moreover, using consumer panels, other authors confirmed the presence of similar attributes in studies with Hass avocado fruit, and in other cultivars reported as Hass hybrids [18,34]. In the aforementioned work, sensory attribute scaled were limited to a creamy texture (watery to creamy), rich (bland to rich), and grassy flavor (grassy to not grassy) on 15-cm line scales. It was not clear if the 'richness' definition was evaluated as an overall attribute or if it was defined for flavor or texture, but it was clearly correlated to the creamy texture attribute (r = 0.86) [15].

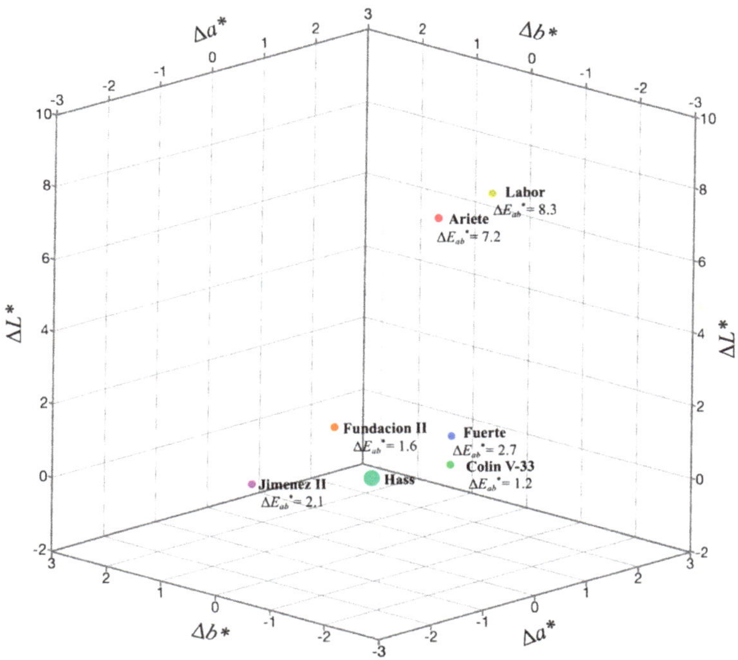

Figure 5. Instrumental colorimetric differences for commercial and non-commercial avocado cultivars in reference to the widely accepted Hass cultivar. Instrumental data for the Hass cultivar were used as reference control to calculate the deviations of each cultivar for the colorimetric parameters, which included ΔL^*, Δa^*, Δb^*, and total color difference (ΔE_{ab}^*).

Literature on sensory studies conducted with avocado cultivars other than Hass and its hybrids was found to be very scarce [16,17]; and as previously mentioned, works conducted with trained descriptive panels on the evaluation of different avocado cultivars were not found. Among the few works that included various cultivars, Shaw et al. [17] conducted sensory hedonics on twenty-one avocado samples; many of them belonged to the West-Indian avocado race characterized by having lower oil contents but described as being well adapted to subtropical regions. West-Indian hybrids are therefore commercially grown in Florida, USA [35]. Consumers that evaluated the pulps documented flavor sensory descriptors, which included nutty, sweet, bitter, and mild. Formal descriptive sensory analyses with trained panels were published for the Hass cultivar samples [8,36]. However, the aims of both studies were different from the identification of drivers of liking and focused on documenting the effects of emerging technologies and storage on sensory profiles. Salgado-Cervantes et al. [36] using an experienced sensory panel identified twelve descriptors that were classified into visual appearance (homogeneity, shiny, and color), aroma (avocado, boiled vegetable, and nutty), flavor (bitter, fatty, and astringent), and texture (unctuous, grainy texture, and fibrous). The attributes were reported to be relevant sensory descriptors present in the control and flash vacuum-expansion processed avocado samples.

4.3. Preference Mapping of Consumer Acceptability and Descriptive Sensory Attributes

Most avocado fruits grown commercially in the world are from the Hass cultivar, since consumers are positively drawn to its taste and texture [18]. Our consumer acceptability results confirmed that Hass presented high liking scores (Table 5). Unexpectable high liking scores were also obtained for non-commercial cultivar Colin V-33, which showed non-significantly different liking scores when compared to Hass. Our results were in agreement with observations reported by López-López [37]. In their work, the authors conducted a small consumer acceptability study (n = 14) using three of the same cultivars evaluated herein (Hass, Fuerte, and Colin V-33). Their results indicated that hedonic scores for flavor, color, odor, and external appearance for cultivar Colin V-33 were not significantly different from those of Hass. Therefore, the observations from both independent studies confirmed that Colin V-33, a non-commercial cultivar, was highly liked by consumers.

In the present study, the external preference map was modeled using overall liking scores from consumers, followed by its placement over a previously constructed descriptive map (Figure 3). Overall liking scores were highly correlated to the appearance, texture, and flavor liking scores (r = 0.87, 0.96, and 0.96, respectively, $p < 0.05$). It is possible that consumers being untrained assessors were not able to accurately differentiate appearance, flavor, and texture but they clearly indicated that the three attributes were relevant for consumer satisfaction of avocado fruit. Pereira et al. [16] also observed for avocado samples that it is common in consumer research to observe correlations among scores for overall liking with those for the liking of specific attributes; possibly because of the halo effect, since consumers are more focused on the general affective response and when they like or dislike a sample, they tend to give similar scores for all its attributes. Obenland et al. [15] conducted a follow-up of the changes in sensory attributes and hedonics during maturation of Hass cultivar from different locations and harvesting years. The aforementioned study showed that liking declined when the texture descriptor for creaminess declined and the flavor descriptor for grassiness increased, indicating that both flavor and texture attributes contributed to liking. Our results confirmed and complemented these observations, since flavor and texture sensory attributes were found to be the drivers of liking; but more specifically flavor descriptors such as lipidic notes, sweetness, and some fresh/green notes, together with texture descriptors for firmness, creaminess, and lipidic residue. The present work also characterized bitter and astringent as sensory attributes present in the avocado pulps, in agreement with descriptive work on avocados conducted by Salgado-Cervantes et al. [36]. Statistical comparisons among pulps did not show marked differences for bitterness or astringency (Table 4), but in the preference map (Figures 2 and 3) both attributes were associated with pulps that were penalized in liking (Labor and Fuerte). Sensory flavor is complex, and for the present work it was limited to the studied attributes, therefore it is possible that those particular pulps transmitted sensations that require further descriptive work.

4.4. Proximate Macronutrient, Instrumental Texture, Instrumental Color, and Sensory Relationships

As previously discussed, total lipids are widely reported in the literature to be a desirable quality in avocado fruit [30]. In this work, all studied cultivars were Guatemalan X Mexican hybrids, and their proximate composition indicated that all were above the California Avocado Industry standard for minimum dry matter percentage (20.8% set for Hass) [30]. However, in the present work we were able to observe that a balance between carbohydrates, sugars, and lipids appears to be relevant to avocado sensory flavor profile (Figure 4). An observation that was also supported by direct slight correlations between affective flavor liking with carbohydrate and sugar contents of the pulps ($r = 0.65$ and 0.61, respectively).

In their work with Hass cultivar, Obenland et al. [15] concluded that carbohydrates, because of their low concentrations, might not influence acceptability. However, their observations could be limited by the use of that single cultivar. A prior study conducted with various cultivars, including different avocado races, focused on carbohydrates [17], and showed that the West-Indian cultivars contained higher levels of seven carbon (C7) sugars, which are rare in nature but are present in avocado fruit. The C7 sugars D-mannoheptulose and perseitol were the main sugars present in some cultivars of the West-Indian race background. Furthermore, West-Indian race fruits contained higher concentrations of the C7 sugars than those for glucose and fructose, and the consumer panel associated them with the sweet sensory attribute [17]. In the present work, the concentrations of individual C7 sugars were not measured; therefore, we were not able to confirm prior author conclusions that when the Mexican race was present in the genetic background, the C7 sugar levels tended to be low [17]. In this study, results from total sugars concentrations were significantly higher for some of the pulps with higher liking scores (Hass and Jimenez II), however, Colin V-33 had lower sugar levels but a higher sweetness sensory scores. Perhaps further work on the characterization of individual sugar profiles, including C7 sugars, can provide further insight into the sensory observations. Flavor metabolites were also reported to play relevant roles in the generation of desirables profiles, and they can be generated from both lipids and carbohydrates, particularly sugars. Lipid degradation products such as acetaldehyde, methyl acetate, 2,4 heptadienal were associated to have high preference values [15]; nonetheless, avocado sugar metabolites are least known. Prior studies described the disappearance of C7 sugars during ripening [1,38] and suggested a potential role in the generation of flavor metabolites.

In addition to flavor, sensory texture and appearance attributes need to be considered among the potential drivers of liking of avocado pulp. Thus, in this study, sensory texture attributes clearly differentiated avocado pulps and were related to the descriptors of lipidic oral sensations. Similarly, other studies confirmed the relationship between texture attributes such as firm, creamy, smooth, and high hedonic scores [15,34]. Herein, instrumental texture measurements alone were poorly correlated to consumer liking, possibly because liking is a complex variable and is difficult to relate to individual instrumental parameters. Nevertheless, instrumental data were useful as an additional objective assessment of the characteristics of the avocado pulps evaluated. Sensory firmness was directly related to instrumental weight of shear ($r = 0.64$), and inversely to instrumental stickiness ($r = -0.69$). Similarly, sensory cohesiveness (rated higher in the most liked pulps) was also inversely related to instrumental stickiness ($r = -0.76$). These correlations served to reassure that the train panel assessments were in accordance with the texture lexicon definitions, since other authors reported similar sensory and instrumental texture relationships for semi-solid matrixes [39].

Interestingly, correlations between sensory texture attributes evaluated by a trained panel, and proximate compositions were even stronger than those for instrumental texture measurements. For instance, sensory cohesiveness showed a significant ($p = 0.0002$) and strong inverse correlation with moisture contents ($r = -0.97$). Additionally, sensory cohesiveness was strongly correlated with total carbohydrates ($r = 0.96$) and sugars ($r = 0.95$), and mildly correlated with lipids ($r = 0.86$), confirming the relevant relationship of both macronutrients to texture, in addition to flavor. Data from both sensory assessments (liking and descriptive) indicated that high moisture levels, thus lower dry matter, lipids, carbohydrates, and sugars moved the texture away from the desired sensations. Contrary to sensory cohesiveness, instrumental stickiness loaded in the same PCA quadrant of the less desirable traits (Figure 4) and was also found to be directly correlated to moisture ($r = 0.88$) and inversely to lipid content ($r = -0.79$). Therefore, results indicated that stickiness was considered as an interesting instrumental texture parameter, since it was also described by prior authors as an undesirable trait for semi-solid matrixes, such as fat spreads [40].

Considering lineage information shown in Table 1, it was observed that non-commercial cultivars that were located in the high-acceptability regions (Colin V-33 and Jimenez II) had common lineages with at least one commercial cultivar (Hass or Fuerte); therefore, suggesting that the progenitors were already selected for the desirable traits, such as flavor and texture. Selection was possibly performed considering high dry matter and oil contents since both chemical traits are known to drive acceptability [33]. Energy concentrations (kcal/100 g fresh weight (FW)), which served as a combined measurement of the contribution of lipids, carbohydrates, and proteins to the overall composition of the pulps are also included in Table 3, and the results clearly indicated that the most liked cultivars contained the highest values (201.3–267.2 kcal/100 g FW). However, results obtained for the Fuerte cultivar indicated that other factors could also influence liking. The Fuerte cultivar ranked in the second-best group for overall liking (Table 5 and Figure S1); although its dry matter, lipid, and caloric contents were the highest of all (Table 3). Sensory texture characteristics of Fuerte were not very different from those of Hass, Jimenez II, and Colin V-33, although the instrumental texture parameters indicated significantly higher values for its work of shear, work of adhesion, and stickiness (Table 6). It is also possible that its acceptability was slightly penalized because of its visual aspects, since its liking of appearance score by consumers were significantly lower (5.9 in a nine-levels hedonic scale, Table 5). Using Hass as a reference, the Δa^* and Δb^* values were similar for Fuerte and the non-commercial cultivars, but the ΔE_{ab}^* values showed some differences among the pulps (Figure 5). Labor ($\Delta E_{ab}^* = 8.31$) and Ariete ($\Delta E_{ab}^* = 7.25$) pulps showed the highest color differences, while Fuerte ($\Delta E_{ab}^* = 2.7$) color variation was not as high. However, Fuerte was slightly different from Hass compared to the more liked Colin-V33 ($\Delta E_{ab}^* = 1.2$) and Jimenez II ($\Delta E_{ab}^* = 2.1$). Perhaps that slight color difference was sufficient to penalize the liking of Fuerte for appearance. However, Ghidouche et al. [24] observed that ΔE_{ab}^*values greater than 3.5 units were required to perceive a color difference by an average observer. Labor and Ariete pulps presented ΔE_{ab}^*values greater than 3.5 from Hass (8.3 and 7.25, respectively), which might have partly influenced consumers' slight overall dislike.

5. Conclusions

For the first time, the development of highly detailed descriptive profiles of different commercial and non-commercial avocado cultivars generated new knowledge on key sensory attributes that drove the liking for avocado pulp conveyed by consumers. Our results confirmed observations obtained from prior consumer evaluations, in which flavor and texture sensory attributes were concluded to be key for liking. Furthermore, the present study's sensory-driven strategy generated an external preference map that facilitated the identification of sensory descriptors, which influenced the overall liking. In general, consumers tend to prefer avocados with a strong global impact, a creamy and oily flavor attributes, and other relevant sensory texture attributes that grouped in Hass's region (one of the most preferred pulps). A non-commercial cultivar, Colin V-33, presented sweet and green notes that also appear to drive preference. Therefore, the results indicated that the drivers of liking for avocado pulp include specific lipid flavor notes, sweetness, green notes, and textures of creaminess/oiliness, lipid residue, firmness, and cohesiveness. The earthy, bitter notes, absence of fibers, and a balanced green color also complemented specific cultivars' preferences. The role of avocado sugars in flavor remains to be further explored since the fruit contains unique carbohydrates. The present work also generated new knowledge and ideas on the possible drivers of disliking, such as stickiness, differences in color, and possibly other unexplored flavors and chemical sensations that remain to be characterized. However, results from the preference map generated valuable information that can be used by avocado breeders and processors as sensory-guided insight to develop and select cultivars with high acceptability for their commercialization strategies.

Supplementary Materials: The following are available online at https://www.mdpi.com/2304-8158/10/1/99/s1, Figure S1. Bidimensional external preference map obtained by quadratic modeling of the overall liking of frequent avocado consumers ($n = 116$) for seven avocado cultivars, and placement of affective data within the sensory descriptive space. Table S1. Instrumental colorimetric values of seven commercial and non-commercial avocado cultivar. Table S2. Differences in instrumental colorimetric parameters of commercial and non-commercial avocado cultivars in reference to the values obtained for cultivar Hass.

Author Contributions: Conceptualization, R.V.-L., D.G.R.-S., M.d.l.C.E.B., R.I.D.d.l.G., and C.H.-B.; Methodology, L.M.M.-O., R.V.-L., D.G.R.-S., M.d.l.C.E.B., A.D.F.-M., J.S.J.-D.l.G., R.I.D.d.l.G., and C.H.-B.; Conceptualization, C.H.-B., A.D.F.-M., R.V.-L., R.I.D.d.l.G., and D.G.R.-S. Software, L.M.M.-O., R.V.-L., and C.H.-B.; Validation, R.V.-L., D.G.R.-S., and C.H.-B.; Formal analysis, L.M.M.-O., R.V.-L., and J.S.J.-D.l.G.; Investigation, L.M.M.-O., R.V.-L., and D.G.R.-S.; Resources, M.d.l.C.E.B.; Data curation, L.M.M.-O., R.V.-L., J.S.J.-D.l.G., R.I.D.d.l.G., and C.H.-B.; Writing—original draft preparation, L.M.M.-O. and C.H.-B.; Writing—review and editing, L.M.M.-O., R.V.-L., D.G.R.-S., M.d.l.C.E.B., A.D.F.-M., R.I.D.d.l.G., and C.H.-B.; Visualization, L.M.M.-O., R.V.-L., D.G.R.-S., and C.H.-B.; Supervision, D.G.R.-S.; Project administration, D.G.R.-S. and C.H.-B.; Funding acquisition, R.V.-L. and C.H.-B. All authors have read and agreed to the published version of the manuscript.

Funding: This research was funded by Tecnologico de Monterrey Translational-Omics (GIEE) Research Group and by the Mexican National Council for Research and Technology (CONACyT) doctoral scholarships for Luis Martín Marín-Obispo (No. 445941), Raul Villarreal-Lara (No. 359813), and Salvador Jaramillo-De la Garza (No. 743316). Descriptive tests were kindly funded by SensoLab Solutions, SC through the SensoLab Pro Bono Research Initiative.

Institutional Review Board Statement: The study was conducted according to the guidelines of the Declaration of Helsinki and approved by the Institutional Ethics Committee for Sensory Evaluation Studies of the Department of Bioengineering, School of Engineering and Sciences of Tecnológico de Monterrey, Campus Monterrey, Mexico (Ethics ID: CSERDBT-0001, approved on 02/12/2018).

Informed Consent Statement: Informed consent was obtained from all subjects involved in the study.

Data Availability Statement: Data is contained within the article or supplementary material.

Acknowledgments: We are grateful to Fundacion Salvador Sanchez Colin-CICTAMEX for providing the avocado cultivars used in this study. A special acknowledgment to CIT2, the Technology Park and Technology Transfer Center for technology-based startups and spinoffs from Tecnologico de Monterrey, for access to their facilities.

Conflicts of Interest: The authors declare no conflict of interest.

References

1. Pedreschi, R.; Uarrota, V.; Fuentealba, C.; Alvaro, J.E.; Olmedo, P.; Defilippi, B.G.; Meneses, C.; Campos-Vargas, R. Primary metabolism in avocado fruit. *Front. Plant Sci.* **2019**, *10*, 795. [CrossRef] [PubMed]
2. Rendón-Anaya, M.; Ibarra-Laclette, E.; Méndez-Bravo, A.; Lan, T.; Zheng, C.; Carretero-Paulet, L.; Perez-Torres, C.A.; Chacón-López, A.; Hernandez-Guzmán, G.; Chang, T.H.; et al. The avocado genome informs deep angiosperm phylogeny, highlights introgressive hybridization, and reveals pathogen-influenced gene space adaptation. *Proc. Natl. Acad. Sci. USA* **2019**, *116*, 17081–17089. [CrossRef] [PubMed]
3. FAOSTAT Avocado: Production-Crops. Available online: http://www.fao.org/faostat/en/ (accessed on 22 September 2020).
4. Dreher, M.L.; Davenport, A.J. Hass Avocado Composition and Potential Health Effects. *Crit. Rev. Food Sci. Nutr.* **2013**, *53*, 738–750. [CrossRef] [PubMed]
5. Duque, A.M.R.; Londoño-Londoño, J.; Álvarez, D.G.; Paz, Y.B.; Salazar, B.L.C. Comparison of the oil from Hass variety avocado cultivated in Colombia, obtained by supercritical fluids and by conventional methods: A perspective under quality terms. *Rev. Lasallista Investig.* **2013**, *9*, 151–161.
6. Farines, M.; Soulier, J.; Rancurel, A.; Montaudoin, M.G.; Leborgne, L. Influence of avocado oil processing on the nature of some unsaponifiable constituents. *J. Am. Oil Chem. Soc.* **1995**, *72*, 473–476. [CrossRef]
7. Ding, H.; Chin, Y.W.; Kinghorn, A.D.; D'Ambrosio, S.M. Chemopreventive characteristics of avocado fruit. *Semin. Cancer Biol.* **2007**, *17*, 386–394. [CrossRef]
8. Jacobo-Velázquez, D.A.; Hernández-Brenes, C. Stability of avocado paste carotenoids as affected by high hydrostatic pressure processing and storage. *Innov. Food Sci. Emerg. Technol.* **2012**, *16*, 121–128. [CrossRef]

9. Lu, Q.Y.; Zhang, Y.; Wang, Y.; Wang, D.; Lee, R.P.; Gao, K.; Byrns, R.; Heber, D. California hass avocado: Profiling of carotenoids, tocopherol, fatty acid, and fat content during maturation and from different growing areas. *J. Agric. Food Chem.* **2009**, *57*, 10408–10413. [CrossRef]
10. Ozdemir, F.; Topuz, A. Changes in dry matter, oil content and fatty acids composition of avocado during harvesting time and post-harvesting ripening period. *Food Chem.* **2004**, *86*, 79–83. [CrossRef]
11. Rodriguez-Saona, C.; Maynard, D.F.; Phillips, S.; Trumble, J.T. Avocadofurans and their tetrahydrofuran analogues: Comparison of growth inhibitory and insecticidal activity. *J. Agric. Food Chem.* **2000**, *48*, 3642–3645. [CrossRef]
12. USDA Avocados, raw, all commercial varieties. Available online: https://fdc.nal.usda.gov (accessed on 22 September 2020).
13. Bill, M.; Sivakumar, D.; Thompson, A.K.; Korsten, L. Avocado Fruit Quality Management during the Postharvest Supply Chain. *Food Rev. Int.* **2014**, *30*, 169–202. [CrossRef]
14. Kadam, S.; Salunke, D.K. (Eds.) Avocado. In *Handbook of Fruit Science and Technology. Production, Composition, Storage and Processing*; Marcel Dekker: New York, NY, USA, 1995; pp. 363–376. ISBN 0-8247-9643-8.
15. Obenland, D.; Collin, S.; Sievert, J.; Negm, F.; Arpaia, M.L. Influence of maturity and ripening on aroma volatiles and flavor in "Hass" avocado. *Postharvest Biol. Technol.* **2012**, *71*, 41–50. [CrossRef]
16. Pereira, M.E.C.; Sargent, S.A.; Sims, C.A.; Huber, D.J.; Crane, J.H.; Brecht, J.K. Ripening and sensory analysis of Guatemalan-West Indian hybrid avocado following ethylene pretreatment and/or exposure to gaseous or aqueous 1-methylcyclopropene. *Postharvest Biol. Technol.* **2014**. [CrossRef]
17. Shaw, P.E.; Wilson, C.W.; Knight, R.J. High-Performance Liquid Chromatographic Analysis of D-manno-Heptulose, Perseitol, Glucose, and Fructose in Avocado Cultivars. *J. Agric. Food Chem.* **1980**, *28*, 379–382. [CrossRef]
18. Pisani, C.; Ritenour, M.A.; Stover, E.; Plotto, A.; Alessandro, R.; Kuhn, D.N.; Schnell, R.J. Postharvest and sensory evaluation of selected 'hass' × 'bacon' and 'bacon' × 'hass' avocado hybrids grown in East-Central Florida. *HortScience* **2017**, *52*, 880–886. [CrossRef]
19. Alemán-Reyes, J.C.; Espíndola-Barquera, M.C.; Barrientos-Priego, A.; Campos-Rojas, E.; Aguilar-Melchor, J.J.; Zarate Chávez, J.J.; López-Jimenez, A. *Variedades, Selecciones y Variedades Criollas de Uso Común*; Fundacion Salvador Sanchez Colin CICTAMEX S.C.: Estado de Mexico, Mexico, 2009.
20. Espíndola-Barquera, M.C.; Barrientos-Priego, A.; Hernández-Escobar, C.; Campos-Rojas, E.; Ríos-Santos, E.; González-Santos, R. *Diversidad de Aguacate en Mexico*; Fundacion Salvador Sanchez Colin CICTAMEX S.C.: Estado de Mexico, Mexico, 2017.
21. López-López, L.; Barrientos-Priego, A.F.; Ben Ya'acov, A.D. Study of Avocado genetic resources and related kinds species at the Fundacion Salvador Sanchez Colin CICTAMEX S.C. In *Proceedings of the Memoria Fundacion Salvador Sanchez Colin CICTAMEX S.C., Coatepec Harinas, Mexico, 1998–2001*; Fundacion Salvador Sanchez Colin CICTAMEX S.C.: Estado de Mexico, Mexico, 2001, pp. 188–201.
22. Rodríguez-López, C.E.; Hernández-Brenes, C.; Díaz De La Garza, R.I. A targeted metabolomics approach to characterize acetogenin profiles in avocado fruit (Persea americana Mill.). *RSC Adv.* **2015**, *5*, 106019–106029. [CrossRef]
23. AOAC. *Official Methods of Analysis of AOAC International*, 18th ed.; AOAC: Rockville, MD, USA, 2005; ISBN 0935584544.
24. Ghidouche, S.; Rey, B.; Michel, M.; Galaffu, N. A Rapid tool for the stability assessment of natural food colours. *Food Chem.* **2013**, *139*, 978–985. [CrossRef]
25. Muñoz, A.M.; Civille, G.V. Universal, product and attribute specific scaling and the development of common lexicons in descriptive analysis. *J. Sens. Stud.* **1998**, *13*, 57–75. [CrossRef]
26. Higa, F.; Koppel, K.; Chambers, E. Effect of Additional Information on Consumer Acceptance: An Example with Pomegranate Juice and Green Tea Blends. *Beverages* **2017**, *3*, 30. [CrossRef]
27. Meullenet, J.F.; Lovely, C.; Threlfall, R.; Morris, J.R.; Striegler, R.K. An ideal point density plot method for determining an optimal sensory profile for Muscadine grape juice. *Food Qual. Prefer.* **2008**, *19*, 210–219. [CrossRef]
28. Viejo, C.G.; Villarreal-Lara, R.; Torrico, D.D.; Rodríguez-Velazco, Y.G.; Escobedo-Avellaneda, Z.; Ramos-Parra, P.A.; Mandal, R.; Singh, A.P.; Hernández-Brenes, C.; Fuentes, S. Beer and consumer response using biometrics: Associations assessment of beer compounds and elicited emotions. *Foods* **2020**, *9*, 821. [CrossRef] [PubMed]
29. Cajuste-Bontemps, J.F.; López-López, L.; Meza-Rangel, J. Fruit quality evolution of two avocado selections (aries and ariete) through their physical attributes. In *Memoria Centro de Investigaciones Científicas y Tecnológicas del Aguacate en el Estado de México.*; Fundacion Salvador Sanchez Colin CICTAMEX S.C.: Estado de Mexico, Mexico, 2000; pp. 37–45.
30. Lee, S.K.; Young, R.E.; Schiffman, P.M.; Coggins, C.W. Maturity Studies of Avocado Fruit Based on Picking Dates and Dry Weight. *J. Am. Soc. Hortic. Sci.* **1983**, *108*, 390–394.
31. Yahia, E.M.; Woolf, A.B. Avocado (Persea americana Mill.). In *Postharvest Biology and Technology of Tropical and Subtropical Fruits*; Yahia, E.M., Ed.; Woodhead Publishing Limited: Shaston, UK, 2011; Volume 2, pp. 125–186. ISBN 78-1-84569-734-1.
32. Carvalho, C.P.; Velásquez, M.A.; Van Rooyen, Z. Determinación del índice mínimo de materia seca para la óptima cosecha de aguacate 'Hass' en Colombia. *Agron. Colomb.* **2014**, *32*, 399–406. [CrossRef]
33. Hodgkin, G.B. *Avocado Standardization California Avocado Society Yearbook*; California Avocado Society: Escondido, CA, USA, 1939.
34. Arpaia, M.L.; Collin, S.; Sievert, J.; Obenland, D. 'Hass' avocado quality as influenced by temperature and ethylene prior to and during final ripening. *Postharvest Biol. Technol.* **2018**, *140*, 76–84. [CrossRef]
35. Bergh, B.; Ellstrand, N. Taxonomy of the Avocado. *Calif. Avocado Soc. Yearb.* **1986**, *70*, 135–146.

36. Salgado-Cervantes, M.; Servent, A.; Maraval, I.; Varga-Ortiz, M.; Pallet, D. Flash Vacuum-Expansion Process: Effect on the Sensory, Color and Texture Attributes of Avocado (Persea americana) Puree. *Plant Foods Hum. Nutr.* **2019**, *74*, 370–375. [CrossRef]
37. López-López, L. Análisis sensorial de selecciones de aguacate (Persea americana Mill). In *Memoria Centro de Investigaciones Científicas y Tecnológicas del Aguacate en el Estado de México*; Fundacion Salvador Sanchez Colin CICTAMEX S.C.: Estado de Mexico, Mexico, 2000; pp. 153–157.
38. Lewis, C.E. The maturity of avocados—a general review. *J. Sci. Food Agric.* **1978**, *29*, 857–866. [CrossRef]
39. Carson, K.; Meullenet, J.F.C.; Reische, D.W. Spectral Stress Strain Analysis and Partial Least Squares Regression to predict sensory texture of yogurt using a compression/penetration instrumental method. *J. Food Sci.* **2002**, *67*, 1224–1228. [CrossRef]
40. Sethi, N.; Balasubramanyam, B.V. Optimization of butter fruit incorporated fat spread using instrumental textural analysis: A response surface methodology. *J. Food Meas. Charact.* **2018**, *12*, 859–866. [CrossRef]

Article

A Multisensor Data Fusion Approach for Predicting Consumer Acceptance of Food Products

Víctor M. Álvarez-Pato [1], Claudia N. Sánchez [1], Julieta Domínguez-Soberanes [2], David E. Méndoza-Pérez [2] and Ramiro Velázquez [1,*]

1 Facultad de Ingeniería, Universidad Panamericana, Aguascalientes 20290, Mexico; valvarez@up.edu.mx (V.M.Á.-P.); cnsanchez@up.edu.mx (C.N.S.)

2 Escuela de Negocios Gastronómicos, Universidad Panamericana, Aguascalientes 20290, Mexico; jdominguez@up.edu.mx (J.D.-S.); demendoza@up.edu.mx (D.E.M.-P.)

* Correspondence: rvelazquez@up.edu.mx; Tel.: +52-449-910-6200

Received: 5 May 2020; Accepted: 4 June 2020; Published: 11 June 2020

Abstract: Sensory experiences play an important role in consumer response, purchase decision, and fidelity towards food products. Consumer studies when launching new food products must incorporate physiological response assessment to be more precise and, thus, increase their chances of success in the market. This paper introduces a novel sensory analysis system that incorporates facial emotion recognition (FER), galvanic skin response (GSR), and cardiac pulse to determine consumer acceptance of food samples. Taste and smell experiments were conducted with 120 participants recording facial images, biometric signals, and reported liking when trying a set of pleasant and unpleasant flavors and odors. Data fusion and analysis by machine learning models allow predicting the acceptance elicited by the samples. Results confirm that FER alone is not sufficient to determine consumers' acceptance. However, when combined with GSR and, to a lesser extent, with pulse signals, acceptance prediction can be improved. This research targets predicting consumer's acceptance without the continuous use of liking scores. In addition, the findings of this work may be used to explore the relationships between facial expressions and physiological reactions for non-rational decision-making when interacting with new food products.

Keywords: consumer acceptance prediction; data fusion; emotion recognition; facial expression recognition; galvanic skin response; machine learning; neural networks; sensory analysis

---

1. Introduction

Consumer response and purchase decision are always uncertain and changing. Nevertheless, there is a consensus that consumer behavior has both psychological and physiological components that have a great influence on consumer choices [1,2].

Among the variety of psychological aspects of consumer behavior, we can find the affective response or the different feelings that a purchase may induce in the buyer: uniqueness, proudness associated with social status, excitement, sense of responsibility, and confidence, among others [3]. The social response or how social groups (family, friends, and society in general) influence consumers and the behavioral response or how personality, demographic origin, and lifestyle may determine purchase choices is also considered part of the consumer's psychological aspects. Readers interested in the relation between emotions and food are directed to [4,5] for comprehensive reviews.

The physiological reactions toward products have recently become of interest for the consumer behavioral studies community. Several attempts have been made to measure physiological reactions accurately and, thus, predict a new product's market performance: heart rate, body temperature, galvanic skin response (GSR), electroencephalography (EEG), visual attention, and facial expressions have been considered as potential hints to determine consumer preference towards a product. Still, recognition of physiological reactions elicited by food products is a novel discipline, and the proper algorithms to interpret them are yet to be developed [6].

Within this context, Viejo et al. evaluated in [7] the EEG, heart rate, temperature, and facial expressions of beer consumers. In [8], He et al. recorded facial expressions of participants exposed to orange and fish odors. Motoki et al. implemented an eye-tracking system to evaluate visual attention elicited by food images [9]. Leitch et al. measured responses to sweeteners in tea through a hedonic scale, an emotion term questionnaire, and facial expressions [10]. Danner et al. reported gauging changes in skin conductance level, skin temperature, heart rate, pulse, and facial expressions of people tasting different juice samples [11]. Similarly, other authors have conducted studies with smoked ham [6] and bitter solutions [12]. A common feature of the aforementioned projects is that they all relied on FaceReader [13], a general purpose and commercially available facial emotion recognition (FER) software.

Other approaches have explored brain activity due to food valuation. Kuhn et al. used functional magnetic resonance imaging (fMRI) to study the effects of viewing and trying chocolate products [14]. Motoki et al. reviewed in [15] the influence of food-extrinsic factors and the mechanisms behind their integration in the human brain.

FER has gained much attention in the field of sensory analysis. Two approaches can be identified in FER [16]: the continuous model and the categorical model. While the former postulates a wide spectrum of different emotions, the latter focuses on a discrete set of basic emotions.

In particular, the categorical model, proposed by Paul Ekman in [17], remains the most popular. Through specialized training, Ekman's method allows identifying face emotions by analyzing certain facial muscular activations. Nevertheless, both manual and automated training processes require many hours, while the facial analysis takes around an hour for each minute of video [18]. This is why a good amount of research has been devoted to finding computer algorithms that are able to outperform the current evaluation of facial expressions.

Deep convolutional neural networks (CNNs) have obtained good results in real-world FER applications [19] besides being robust and reducing variations across faces [16]. Among the different works that rely on CNNs for FER are those of Cai et al. [19], who proposed a new loss function, which tries to maximize the differences between predicted classes in FER-applied CNNs. Zhao et al. implemented in [20] a 3D CNN architecture to learn features in facial images and optical flow sequences. Li et al. [21] used an attention mechanism in CNNs to classify facial expressions even in partially occluded faces by focusing on different regions of a facial image and weighing them according to the level of occlusion they exhibit and how much they contribute to the classification. Wang et al. [22] tried to improve recognition accuracy by focusing on combining multiple weighted regions of facial images. Liong et al. designed a shallow (only two layers) triple stream 3D CNN that was capable of extracting high level features, as well as micro-expressions through determination of optical flow features [23].

In addition to FER, novel types of analyses including recording of emotions and physiological changes elicited in consumers when trying new food products have contributed to achieving a more complete understanding of consumer responses [24].

In general, two different kinds of analyses have contributed to improve sensory evaluation: explicit and implicit.

Explicit analyses involve questionnaires that make use of verbal and non-verbal descriptor terms [25–27]. They have many advantages: they are easy to understand by consumers and relatively fast to decode. Some drawbacks are that results might be cognitively biased [27], and they do not record the consumer experience at the precise moment of tasting the product.

On the other hand, implicit methods measure FER and other physiological changes. For the latter, a comprehensive review that explains the response patterns for certain emotions can be found in [28]. Other implicit methods measure heart rate, skin conductance, skin temperature, and pupil dilation, among other physiological changes, and autonomic nervous system responses [25,26,28].

Though it has been reported that the perception of basic tastes is linked with specific facial movements (for example, sourness with the lips or bitterness with the eyes and forehead) [29], many different variables can affect both FER and physiological changes: how hungry a consumer feels, the type of food tested, and the time elapsed since the beginning of the test. Even in tests as short as 10 s, a consumer may exhibit several facial expressions [30]. Moreover, changes in facial expression are harder to determine when tasting food products, compared to smelling a perfume or watching a video [25], since the jaw movement produced by chewing and the occasional facial occlusion (because of the hand that takes the sample to the mouth) are often the cause of misreading in FER algorithms. These might be some of the reasons why similar studies seem inconclusive [7,31,32].

The present work aims to take a step towards a more reliable prediction of consumer acceptance by means of CNNs and other machine learning algorithms that interpret facial expressions and find potential correlations between biometric sensor measurements, facial analysis, and reported liking.

In this work, a self-developed FER system is introduced since not all studies based on commercial solutions have proven to be successful. Programming our own application provides more flexibility and allows exploring different methods and emotion models. Given that emotions are expressed multi-modally [33] and the fusion of information channels helps to improve predictions [34], we included biometric sensors in the analyses as well. The CNN presented in this paper is comprised of four channels, one for each quadrant of the facial image, as previous findings suggested that multiple networks performed better than an individual ones [16].

The rest of the paper is organized as follows: Section 2 presents the materials and methods used for the implementation of the proposed sensory evaluation system. Section 3 presents the results obtained, while Section 4 poses some discussion. Finally, Section 5 concludes the paper, summarizing the main contributions and giving future work perspectives.

2. Materials and Methods

2.1. Sensory Analysis

2.1.1. Flavor and Odor Sample Description

For the experiments, the following ingredients and percentages were used to prepare sweet gums of five different flavors: glucose (36.5%, Deiman, El Paso, TX, USA), sugar (33.21%, Gelita, Lerma, Mexico), water (23.38%), unflavored gelatin (5.3%, Gelita, Lerma, Mexico), citric acid (1.28%, ENSIGN, Shandong, China), flavor (0.3%), and red color (0.03%, Deiman, USA). For two flavors (clam and cheese), maltodextrin was used instead of sugar.

The procedure for the elaboration of sweet gums is depicted in Figure 1 and is as follows:

1. Let unflavored gelatin dissolve in water (10.6 g/L) for 30 min.
2. Mix sugar and water (11.5 g/L) and heat at 70 °C; add glucose, and increase the temperature up to 108 °C.
3. Add the unflavored gelatin solution, color, and flavor to the mixture at 100 °C, as well as diluted citric acid (1.28 g/L).
4. Finally, cast the mixture in a bed of starch and let it rest for 18 h.

All the resulting sweet gums had similar appearances (Figure 1f) in order to keep volunteers from guessing the sample's taste beforehand. Sweet gums can control the flavor release at the precise moment when the product is tasted, and therefore, the facial expression can be measured at the exact moment that the consumer receives the stimuli. These sweet gums' flavors were determined beforehand in order to provide five different sensory stimuli. We used three flavors considered as pleasant: mint (Deiman, USA), pineapple (Deiman, USA), and strawberry (Deiman, USA), as well as two unpleasant ones: clam (Bell, Northbrook, IL, USA) and Gouda cheese (Bell, USA).

We also prepared a set of odor samples by soaking pieces of cotton in different solutions within a sealed container. Test participants would use a small wooden stick to bring the substance closer to their nose (Figure 2). The odors used for this experiments were: pineapple (Ungerer, Lincoln Park, NJ, USA), mint (Deiman, Horizon City, TX, USA), vinegar (Ungerer, Lincoln Park, NJ, USA), Gouda cheese (Bel, Chicago, IL, USA), and smoke (Castells, Mexico city, Mexico).

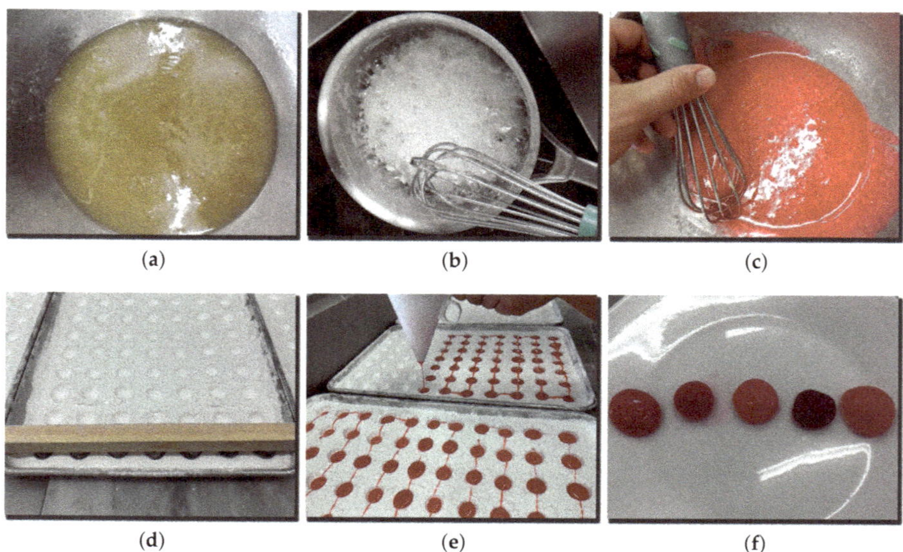

Figure 1. Sweet gum elaboration process: (**a**) gelatin in water, (**b**) sugar, water, and glucose mixture, (**c**) gelatin solution, color, flavor, and citric acid in the mixture, (**d**) bed of starch, (**e**) mixture in the bed of starch, and (**f**) resulting sweet gums.

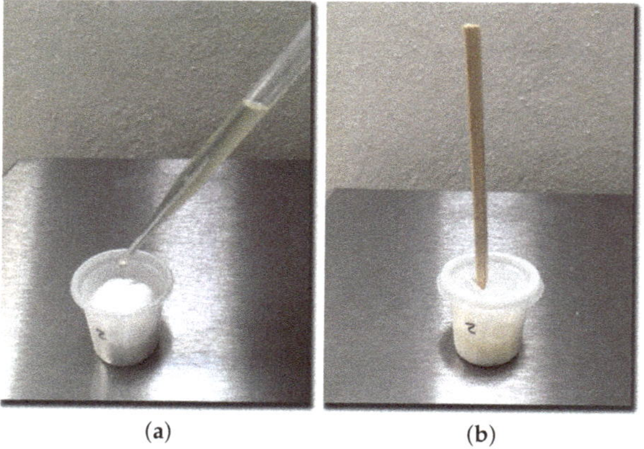

Figure 2. Odor sample: (**a**) Soaking cotton with solution. (**b**) Wooden stick in container.

2.1.2. Participants and Setup

A group of 120 students, professors, and administrative staff from Universidad Panamericana (Mexico) volunteered for the test. The experiment was conducted in a Sensory Laboratory with a controlled illumination booth. The booth was equipped with a Kinect device, which integrated various sensors such as a color camera, an infrared light camera, and a depth sensor, all in a single device, thus eliminating the need to manage and synchronize multiple sources of information.

For the present study, the Kinect device acquired the participants' frontal facial images. During the test, a Neulog NUL-217 device was attached to each volunteer's middle and ring fingers, as well as a Neulog NUL-208 attached to the index finger to measure both galvanic skin response (GSR) and cardiac pulse, respectively. The experimental setup is shown in Figure 3.

A small semaphore device was used to let each participant know the right moment to try each sample. This allowed a better synchronization between the reaction of the user and the records captured by the Kinect camera. All participants were instructed to try a sample at a specific moment while being filmed by the Kinect. After trying each sample, consumers had water and salted crackers to neutralize the flavors. Finally, the consumers were instructed to answer a sensory questionnaire.

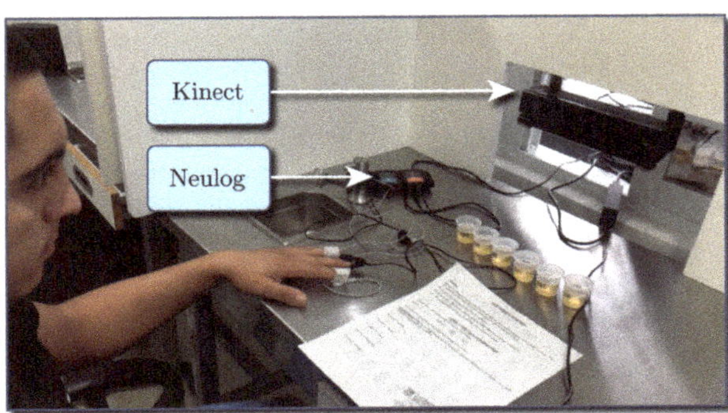

Figure 3. Booth setup.

2.1.3. Questionnaire

A questionnaire comprised of a seven-point hedonic scale for each of the five samples of taste and odor was implemented. These questionnaires are commonly used in sensory science for testing acceptance of different types of food products.

The results obtained with the sensory questionnaires were compared to those obtained from the facial expressions using artificial intelligence methods.

2.2. System Architecture

Figure 4 shows the system's architecture, depicting its main modules. Our sensory analysis system used three inputs: facial images, GSR signals, and cardiac pulse signals. As previously mentioned, facial images were acquired by a Kinect device, while the GRS and cardiac pulse signals via a Neulog device.

Facial images were analyzed by a previously trained CNN to determine the consumer's emotion. The detected emotion together with the GSR and pulse signal values followed a statistical-based data fusion process. The result went to a machine learning model, which in turn predicted the consumer acceptance.

The machine learning model was based on the random forest classification method, and it used the consumer's liking scores and the values resulting from the data fusion stage for training. Once trained, the liking scores were no longer needed. This approach targeted eliminating their use for predicting consumers' acceptance. Having just an implicit method for measuring acceptance eliminated external factors and their influence on the results.

The following subsections will detail the system's modules.

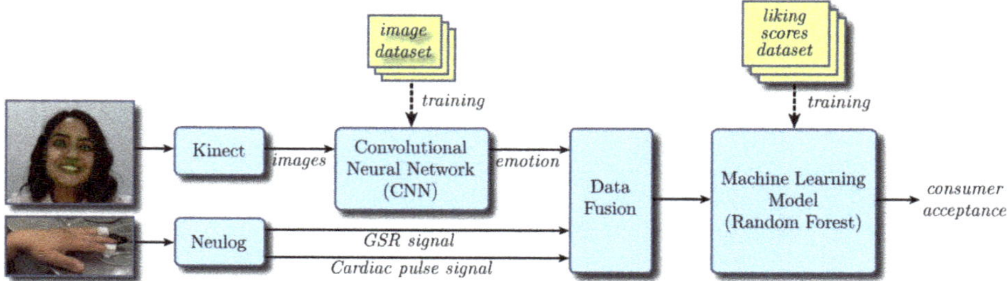

Figure 4. The sensory analysis system architecture. GSR, galvanic skin response.

2.3. Neural Networks

A neural network is an interconnected assembly of processing elements whose functionality loosely resembles that of a neuron [35]. These processing elements are commonly known as perceptrons. Figure 5 shows their basic structure: a set of numerical inputs x_i are weighted by a corresponding factor or weight ω_i. Results are then added. Finally, an activation function σ is applied to the sum to yield the final result y. For any input vector **x** of length n, this functionality can be expressed by Equation (1):

$$y = \sigma(\sum_{i=1}^{n} w_i x_i) \quad (1)$$

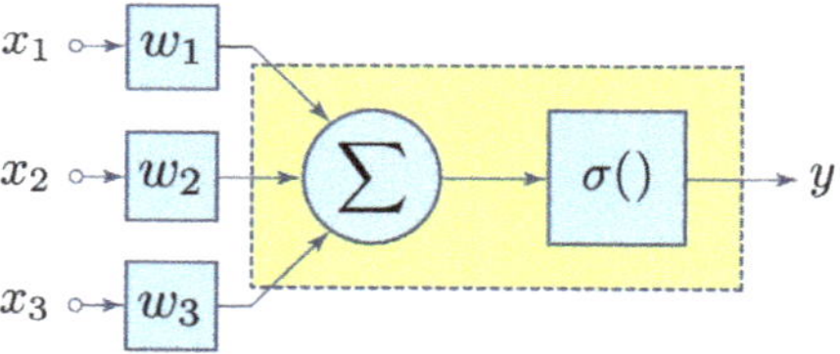

Figure 5. Schematic representation of a three-input perceptron.

Many layers can be stacked together to configure larger neural networks for more complex classification tasks. Figure 6 depicts, for example, a two-layer neural network.

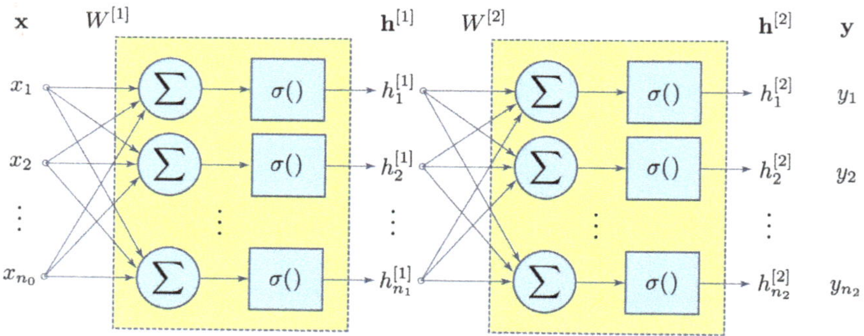

Figure 6. Two-layer neural network architecture. For the sake of simplicity and clarity, individual weights are not shown.

2.4. Facial Expression Datasets

Two different facial expression datasets: AffectNet [36] and CK+ [37], for training and testing the neural network, respectively, were used.

AffectNet contains more than 420,000 facial images classified among 11 discrete labels. However, in order to have a balanced training, only 3800 images associated with each of the next 10 labels were used: neutral, happy, sad, surprise, fear, disgust, anger, contempt, none, and uncertain. Images classified as non-face in the dataset were not included. We chose to use CK+ as the evaluation reference since it is a well-known dataset among researchers and is comprised of a much lower number of labeled images.

2.5. Image Preprocessing

Images in the training dataset had some irregular characteristics that could not be handled by the neural network. Therefore, some preprocessing was required to assure that the network would receive only consistent information, namely:

1. Discard color information, converting RGB-coded images to gray scale in order to reduce their size and the processing time.
2. Detect all faces in the image, together with their bounding rectangles by applying an algorithm based on the histogram of gradients (HoG) [38].
3. Locate 68 landmarks on the first face detected using the Kazemi algorithm [39].
4. Rotate the image and landmarks to make the line between landmarks 40 and 43 horizontal, so that all processed faces are aligned (see Figure 7).
5. Divide facial image into four sections, specifically left and right sections for eyes and nose-mouth.
6. Flip right sections horizontally in order to feed left and right sections to the same network.
7. Equalize every section with contrast limited adaptive histogram equalization (CLAHE) [40].
8. Normalize pixel values in the range from (0,255) to (0,1).

These operations were performed through the Dlib [41] and OpenCV [42] libraries in the Python programming language, while those related to the neural network used the Keras library [43] for Deep Learning with the TensorFlow backend [44].

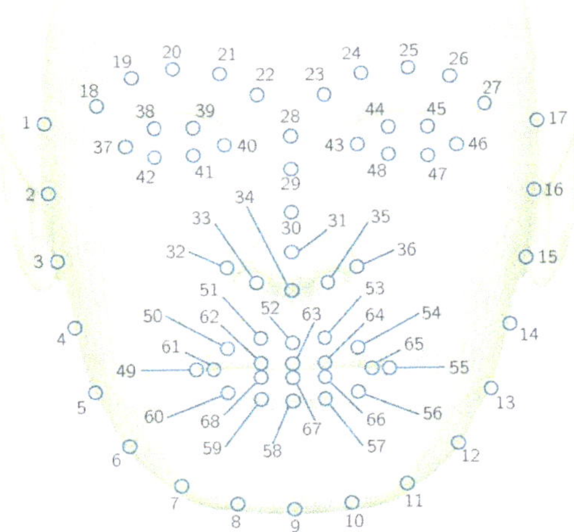

Figure 7. Numbered facial landmarks.

2.6. Network Architecture

The first stage was composed of two networks that were trained differently, but shared the architecture shown in Figure 8: every image section (64 × 64 pixels) was fed through three convolutional layers and a max-pooling layer. Later, the other two similar filter blocks further reduced the 2D information to feed the other four dense row layers, the last of which produced a partial classification into 10 possible labels by means of a softmax transfer function. All previous transfer functions were rectified linear units (ReLUs).

Network A output a 10 number vector for left eye and flipped right eye sections, whereas Network B did the same for nose-mouth sections. The four resulting vectors would then work as a 40 number input for the second stage of the network. This stage encompassed two dense ReLU layers and a softmax output layer that yielded the final classification.

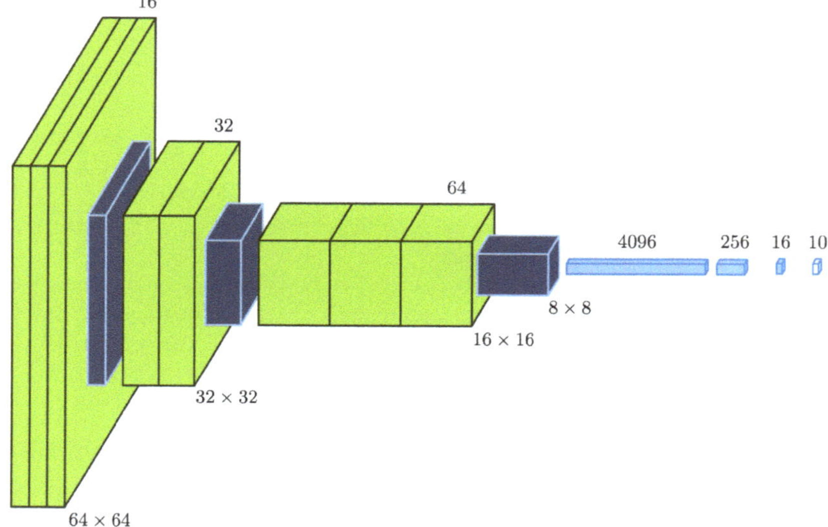

Figure 8. Neural network architecture for the first stage.

2.7. Network Training

Only 40,366 faces of the selected subset were fit for training since the face or landmark detection algorithms did not work properly for all cases. By flipping the right sections, a total of 80,672 facial images were available for training Networks A and B. Both networks were trained on 50 epochs with a batch size of 128. We used 20% of the training set for validation and a dropout rate of 0.4 in some layers to reduce the chance of overfitting.

Next, the networks processed all available faces to obtain 80,672 vectors of 40 elements, which were used as the training set for the second stage. In the corresponding training process, we applied the same parameters as the in first, except for the validation percentage, which was 15%.

2.8 Emotion Recognition

The trained network was fed with all the preprocessed images corresponding to 111 participants. An equal number of CSV files were obtained containing the following columns: image index, number of faces detected (or −1 if the algorithm was unable to find a face), image file name, and the probability of classification for all emotion labels mentioned in Section 2.4.

2.9. Data Fusion

Each an experiment provided data from three different sources: (1) images that provided the response of nine facial expressions in a range (0,1), (2) GSR, and (3) pulse response. As shown in Figure 9, the sensor's measurements were spread across a time series (measured in frames, which were recorded at a target rate of 30 frames per second), and several samples were stored for each experiment. To represent the sensors' data, four statistical metrics were used: the average (avr), the standard deviation (std), the minimum (min), and the maximum (max) values. In sum, for each experiment, there were 44 features obtained from four statistical metrics of nine facial expressions, GSR, and pulse signals.

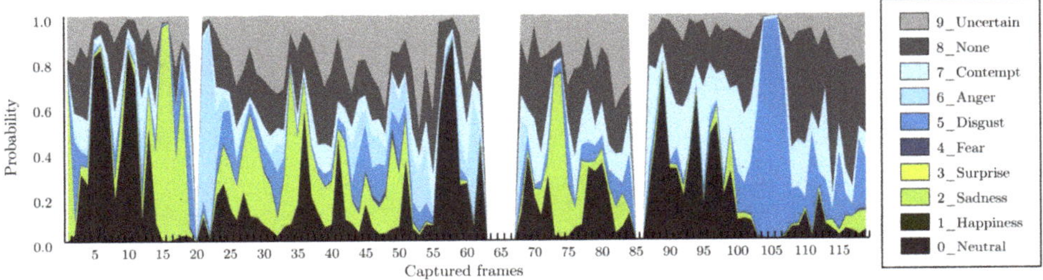

Figure 9. Example of detected emotion probabilities.

2.10. Acceptance Prediction

We used machine learning regression techniques to predict the acceptance that consumers assigned to each sample. For each experiment, we first extracted 44 input features, as previously explained, and one output: the level of acceptance assigned to the sample by the consumer. Each consumer evaluated 10 different samples.

The selected machine learning model for predicting the acceptance was random forest, as proposed by Breiman [45]. It was comprised of a set of random decision trees (30 for this work), each one created with a random subset of samples and features from the training dataset.

A decision tree is a prediction model based on a series of questions about features' values (Figure 10). In a decision tree, data are separated into many dimensions by hyperplanes, which are determined as a result of the questions. The main idea is that samples with similar values tend to concentrate in the same region. We chose random forest because it measures and shows how much each feature contributes to the final model. Decision trees establish selection criteria (Figure 10) by trying to minimize the impurity of the data in each node. In this case, the impurity was calculated as the mean squared error (MSE), formally Equation (2).

$$MSE(\vec{y},\vec{\hat{y}}) = \frac{1}{n}\sum_{i=1}^{n}(y_i - \hat{y}_i)^2, \tag{2}$$

where $\vec{y}$ and $\vec{\hat{y}}$ are the real and the predicted outputs (the reported acceptance values in the experiments), respectively, and n is the number of samples. When a classification rule is defined, the node data are split into two regions. Several features and values were tested, and the feature-value pair that minimized the impurity was selected as the classification rule.

Feature importance is proportional to the impurity reduction of all nodes related to that feature. The impurity reduction IR in each node j representing a rule can be calculated with Equation (3):

$$IR_j = w_j I_j - (w_{left} I_{left} + w_{right} I_{right}), \tag{3}$$

where *left* and *right* represent the children nodes of node *j*, *I* represents the impurity of each node, and the weights *w* are the samples' proportion in nodes, and they are calculated as the number of samples in the node divided by the total number of samples. Once the impurity reduction in all nodes is known, the importance of the feature *k*, FI_k, is calculated using Equation (4):

$$FI_k = \frac{\sum_{j \in N_k} IR_j}{\sum_{j \in N} IR_j}. \quad (4)$$

where N_k represents the set of all nodes that are split using variable *j* and *N* represents all the nodes in the decision tree.

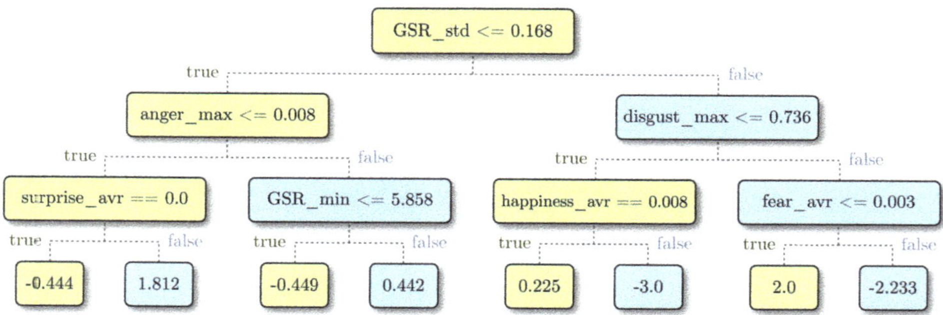

Figure 10. Decision tree example: to predict the consumer's evaluation, questions need to be answered from top to bottom and following the path of the answers. At the end of the path, the last node contains the prediction of the consumer's evaluation.

Results were validated through a ten-fold cross-validation. This meant that the dataset was randomly divided into ten blocks. The model was later fitted ten times using nine blocks for training and one block for testing. The mean absolute error (MAE) was used for calculating the model error. Formally, Equation (5):

$$MAE(\vec{y}, \vec{\hat{y}}) = \frac{1}{n} \sum_{i=1}^{n} |y_i - \hat{y}_i| \quad (5)$$

where $\vec{y}$ and $\vec{\hat{y}}$ are the real and the predicted outputs, respectively. The MAE was calculated each time the model was fitted and tested. The final results were the average of all runs. We chose to present these results with MAE as opposed to the MSE used for fitting the model, since it was easier for interpretation.

3. Results

Figures 11 and 12 show the cumulative results for hedonic scales in the taste and smell evaluations, respectively. The bars, centered on zero, represent how many participants rated each smell or flavor. These figures show the results of the questionnaire with Likert scales ranging from –3 (the most disliked) to three (the most liked).

Strawberry was the most liked flavor, and Gouda cheese seemed to elicit the worst reaction, since its most common score was -3 and almost the whole bar lied on the left side of the chart. Clam obtained a negative overall score as well. As for smell tests, pineapple and mint had a good acceptance, while Gouda cheese, vinegar, and smoke did not. There seemed to be a good contrast between reported acceptance of liked and disliked samples.

Figure 13 shows the emotions that were recognized during the taste (Figure 13a) and smell (Figure 13a) experiments. The boxplots represent the average probability value of each emotion for all consumers during the five experiments. It can be observed that sadness was the emotion that mostly appeared during the execution of the experiments followed by disgust.

Our results agreed with those of He et al. [30]. In this study, He and coworkers measured the changes of facial expression for the same, similar, and different taste conditions. They concluded that the pleasantness of consuming a food product diminished very rapidly; therefore, in this finding, the expressions of sadness and angry were found to be predominant. Additionally, we noted that the expressions of sadness and disgust were probably due to consumers arriving nervous and with uncertain expectations for the experiment, then, once in the test, with their important concentration for perceiving all tastes and odors.

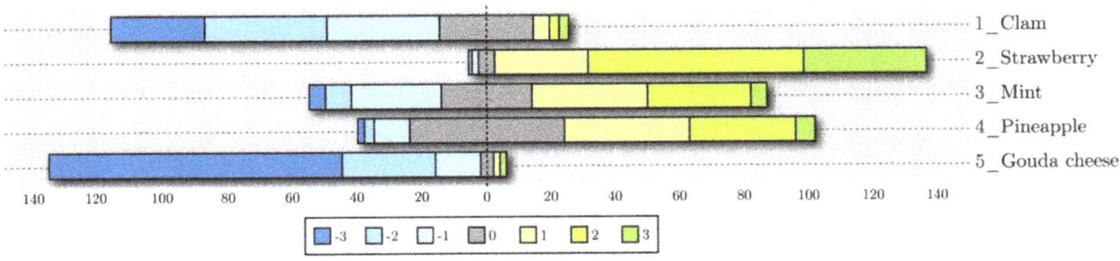

Figure 11. Acceptance results for taste evaluations.

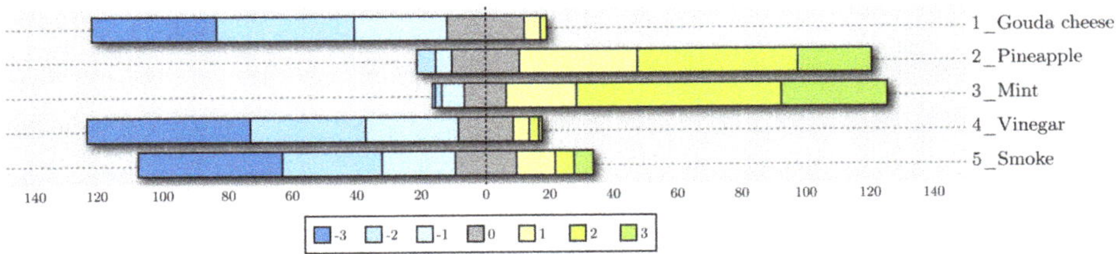

Figure 12. Acceptance results for smell evaluations.

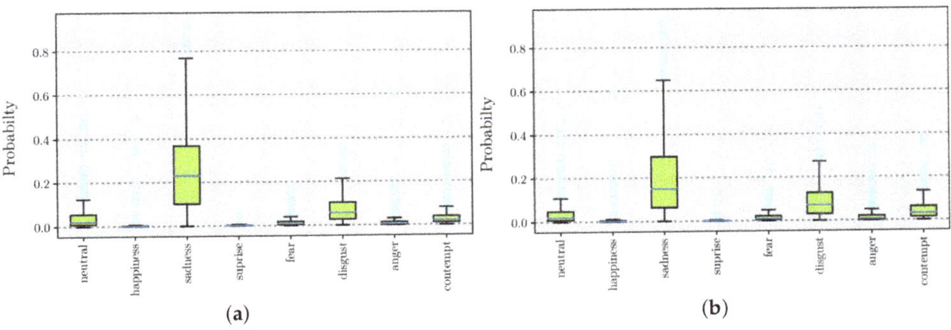

Figure 13. Recognized emotions: (**a**) taste and (**b**) smell experiments.

Figure 14 shows the correlation matrices of FER, sensor responses, and consumer acceptance in the different experiments. Values on the matrices were calculated using the absolute Pearson's correlation coefficient. No strong correlation between consumer acceptance and other features was found. However, the features with higher correlations with consumer acceptance were the following: in Figure 14a: fear, happiness, disgust, pulse, and GSR; in Figure 14b: neutral and happiness; in Figure 14c: GSR, happiness, and disgust; in Figure 14d: disgust and neutral.

We interpreted that fear, happiness, disgust, neutral, pulse, and GSR were the features with higher correlation with consumer acceptance. However, these correlations were too weak. Fear was the most difficult expression to recognize accurately with static images [46]. Fear was usually miscategorized together with surprise by both humans and FER models [46,47].

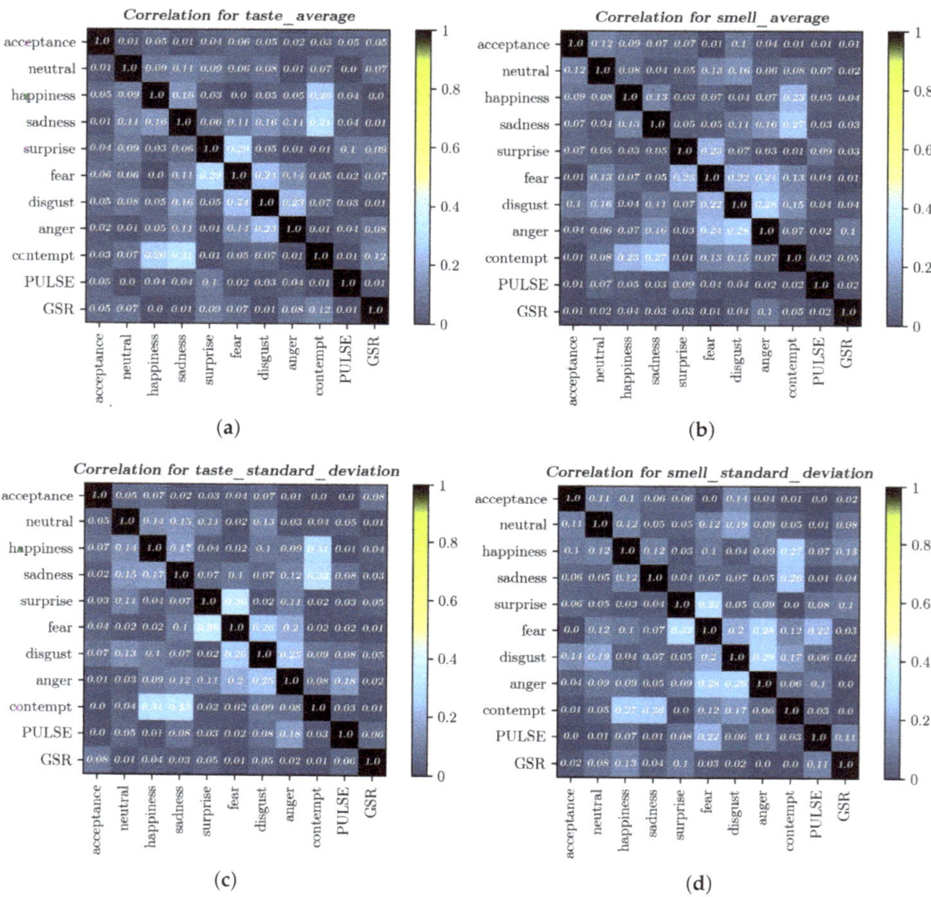

Figure 14. Correlation matrix of facial emotion recognition (FER), sensor responses, and consumer acceptance: (**a**,**c**) display the correlation matrices of taste experiments, while (**b**,**d**) show the correlation matrices of smell experiments. Cells contain correlations between column and row features.

Table 1 shows the MAE of our regression model, as described in Equation (5), which predicted sample acceptance based on FER and the sensors' detected responses. The first column describes the type of data used to train the random forest. The model obtained the best prediction when trained with the GSR measurements alone. These results were similar to those obtained in previous works [48,49].

Table 1. Mean absolute error (MAE) for the regression model.

Data	Taste	Smell
sensors + emotions	1.8216	1.8593
emotions only	1.8408	1.8273
sensors only	1.7896	1.8493
GSR only	1.7649	1.7817
pulse only	1.8173	1.9655

As mentioned in Section 2.10, our random forest model rated the importance of each feature in predicting acceptance. The ten most important features for each set of experiments are shown in Figure 15.

The standard deviations of pulse and GSR samples showed up as the most relevant variables to take into account when predicting acceptance. The average of surprise measurements emerged as the top variable in the left column. However, it was absent in the right one. This could be explained by the fact that the sense of smell may elicit greater emotions than taste. Nevertheless, emotion measurements were very similar in both smell and taste experiments (Figure 13). This suggested that the GSR and pulse sensors were better predictors than the CNN array.

A box plot was obtained for every sample. Note that no relevant variance could be found among them. For this reason, only the average recorded values for every detectable emotion were included, as shown in Figure 13.

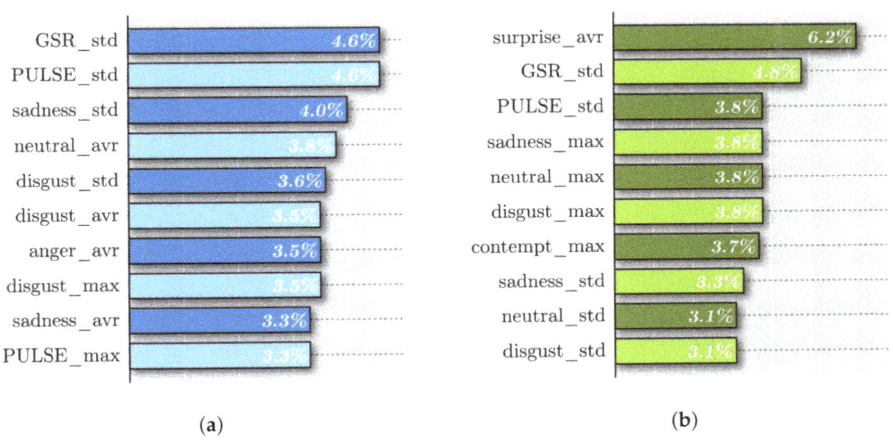

Figure 15. Feature importance: (**a**) taste and (**b**) smell regression models.

We calculated the average value and standard deviation of all participants' GSR and pulse, once for each of the 100 samples, which were obtained at a rate of eight per second in arbitrary units, as provided by the sensors. Figures 16 and 17 display these results: blue graphs represent flavors and smells reported as disliked, whereas those related with liked samples are shown in green. In Figure 16, two flavors seemed to be producing relevant changes in the sensors' measurements, both of which were clearly reported as disliked: the GSR average of samples slowly rose for 1_Clamand abruptly fell for 5_Gouda cheese, while the standard deviation of the latter distinguished itself from the remaining graphs because of its steep increase. No disparity was shown between the taste samples reported as liked: 2_Strawberry, 3_Mint, and 4_Pineapple. Furthermore, they remained almost constant over time.

Figure 17 shows a similar pattern for two curves associated with disliked samples: the value of the GSR average in 5_Smoke plunged, while it stayed above the others for 1_Gouda cheese, with the resulting changes in the standard deviation. 4_Vinegar, on the other hand, followed the same trend as those samples that scored high on liking: 2_Pineapple and 3_Mint. The smell of vinegar might induce weaker reactions than the reported liking scores suggested.

Average pulse readings revealed again characteristic curves for 1 and 5, but only the standard deviation of 5_Smoke showed a clear distinction. Once again, and except for 4_Vinegar, smells reported as disliked were separated from the rest in some way, which suggested that these features were indicative of strong (or at least detectable) emotional reactions.

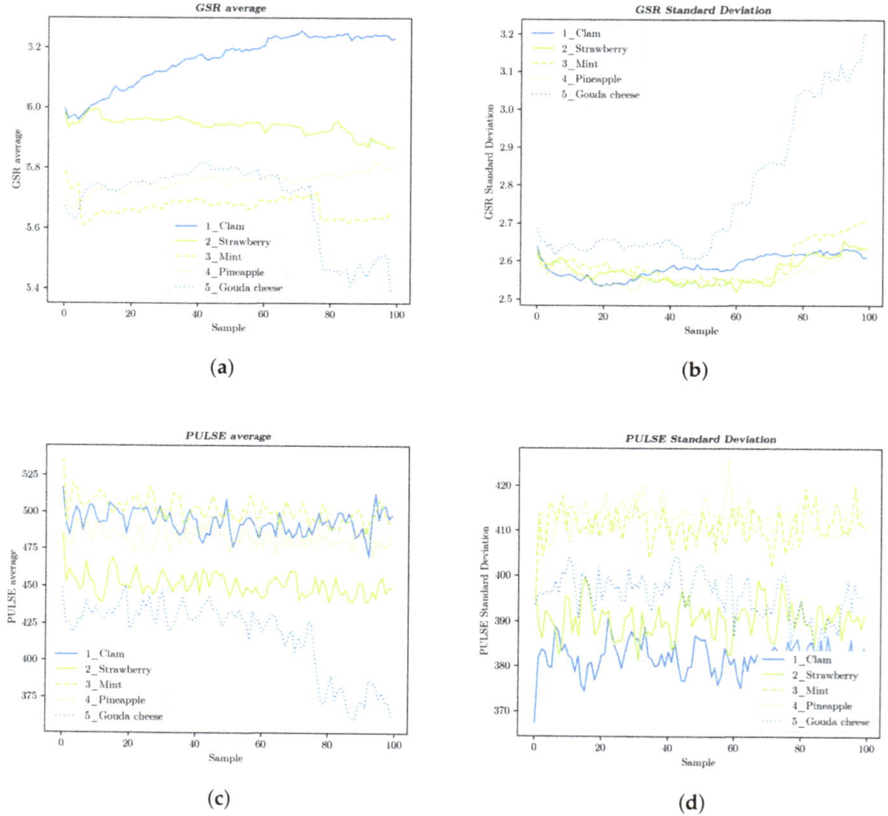

Figure 16. GSR and pulse measurements for the taste tests: (**a**) The participants' GSR average values and (**b**) standard deviation. (**c**) The participants' pulse average values and (**d**) standard deviation.

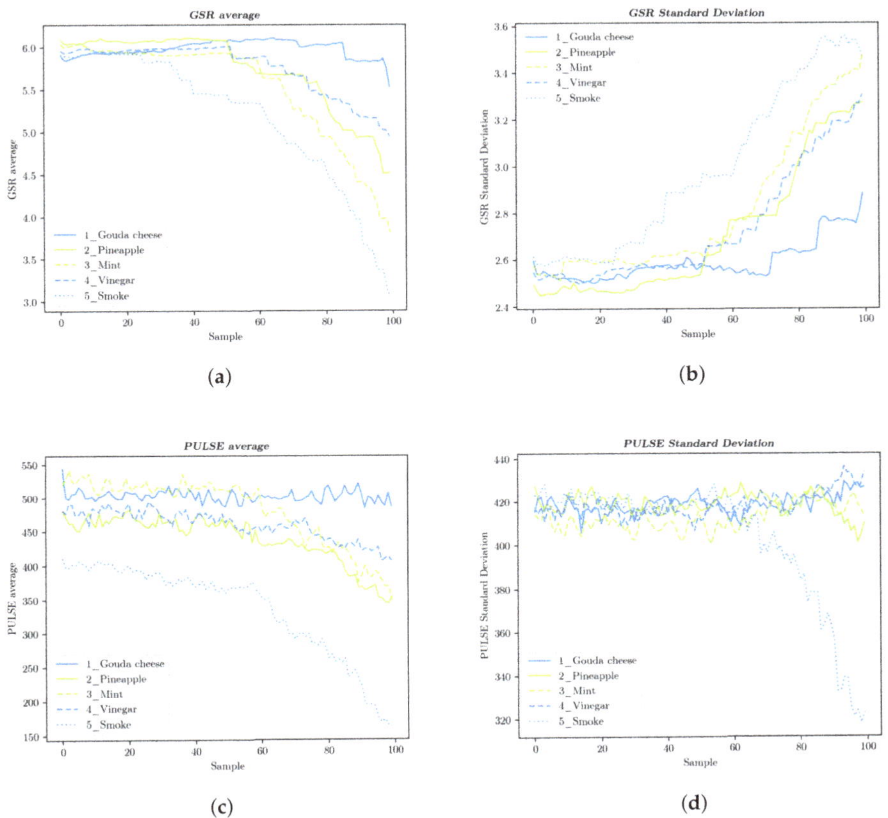

Figure 17. GSR and pulse measurements for the smell tests. (**a**) The participants' GSR average values and (**b**) standard deviation. (**c**) The participants' pulse average values and (**d**) standard deviation.

4. Discussion

Automatic facial emotion recognition (FER) by itself is not a problem with a single straightforward solution; still, many variables must be considered. To this day, the best reference for rating observed emotions is human assessment, which is still prone to misclassifications [33] even with proper training [50], because emotion recognition is context-dependent [33,47] and thus relies heavily on a cognitional understanding of the situation in which an emotion is produced. On top of that, in experiments similar to ours [6,32], consumers showed almost no facial gestures, even for acute stimuli [2], and some expressions proposed as innate were rarely observed [51]. All of this might explain why our FER approach, as well as similar ones constantly detected just a couple of emotions.

However, Bredie et al. [2] and Crist et al. [12] were successful in evoking expressions of disgust with highly concentrated solutions of caffeine, citric acid, and sodium chloride. The use of analogous stimuli should be taken into account for future work as it might help CNNs detect facial expressions more clearly. Gunaratne et al. [24] reported expressions of sadness positively associated with the tasting of salty chocolate. This may provide a hint to find out why the proposed FER system yielded a classification of sadness so often.

Finally, the correlations found between emotions and hedonic scores were very low, just as those reported by Litch [10]. It could be therefore concluded that the connection between food consumption and experienced emotions, as well as the connection between real and FER reported emotions were much weaker than expected, at least when appraised in this manner.

On the other hand, FER for food product assessment has been the subject of very few studies, and the required algorithms are yet to be developed [6]. Nevertheless, our results, as well as those obtained by Samant et al. [26] suggested that GSR measurements were more reliable for pointing out emotional reactions. Though further research is still necessary to confirm that liking can be associated with an increased heart rate as stated by De Wijk et al. [52], their assertion of emotional intensity being associated with a reduced heart rate could be confirmed to some extent by the graphs shown in Figure 17. Further studies should not leave out this type of sensor in order to validate our present results.

5. Conclusions

Measuring physiological signals and images to determine consumer acceptance as part of, or in addition to, other sensory tests is gaining attention in sensory science. In this context, this paper presented a novel automatic sensory analysis system, which aimed to predict consumers' acceptance when trying new food products.

The system encompassed facial emotion recognition (FER), galvanic skin response (GSR), and cardiac pulse together with liking reports. A novel artificial intelligence based approach for data fusion of consumers' facial images and biometric signals was proposed to determine the preference elicited by food products. Two input channels were used and compared: taste and smell.

The experiments conducted to validate this approach involved the participation of a group of 120 voluntary subjects. The significant amount of data obtained was processed using machine learning techniques such as neural networks, statistical metrics, and decision trees.

Results showed that FER alone was not sufficient to determine consumers' acceptance. In line with previously reported works, the facial expressions of sadness and disgust were constantly detected probably due to consumers feeling nervous, anxious, or simply concentrating during the experiment. However, when correlated with GSR and pulse signals, acceptance prediction could be improved. Our experiments showed that GSR was the most relevant variable to take into account when predicting product acceptance. Cardiac pulse, to a lesser extent, could be confirmed to be related to emotional intensity elicited by food products.

The proposed approach was proven to be efficient at processing and correlating different kinds of input signals and big amounts of data. Future studies will investigate the use of EEG signals as an additional biometric input to the model and the use of intense flavors and smells (such as citric acids, perfumes, and hydrogen sulfide) to induce clear facial expressions.

Author Contributions: Conceptualization, C.N.S., J.D.-S., and R.V.; methodology, V.M.Á.-P., C.N.S., D.E.M.-P., J.D.-S., and R.V.; software, V.M.Á.-P. and C.N.S.; validation, V.M.Á.-P. and C.N.S.; formal analysis, C.N.S., J.D.-S., and R.V.; investigation, C.N.S., J.D.-S., and R.V.; resources, J.D.-S.; data curation, V.M.Á.-P. and C.N.S.; writing, original draft preparation, V.M.Á.-P., C.N.S., and J.D.-S.; visualization, V.M.Á.-P. and C.N.S.; supervision, J.D.-S. and R.V.; project administration, J.D.H.; funding acquisition, J.D.-S. All authors read and agreed to the published version of the manuscript.

Funding: This research was funded by Universidad Panamericana through the grant "Fomento a la Investigación UP 2018", under Project Code UP-CI-2018-DNG-AGS.

Conflicts of Interest: The authors declare no conflict of interest.

References

1. Köster, E.P. Diversity in the determinants of food choice: A psychological perspective. *Food Q. Prefer.* **2009**, *20*, 70–82. [CrossRef]
2. Bredie, W.L.; Tan, H.S.G.; Wendin, K. A comparative study on facially expressed emotions in response to basic tastes. *Chemosens. Percept.* **2014**, *7*, 1–9. [CrossRef]

3. Soodan, V.; Pandey, A.C. Influence of emotions on consumer buying behavior: A study on FMCG purchases in Uttarakhand, India. *J. Entrepreneurship Bus. Econ.* **2016**, *4*, 163–181.
4. Macht, M. How emotions affect eating: A five-way model. *Appetite* **2008**, *50*, 1–11. [CrossRef]
5. Evers, C.; Dingemans, A.; Junghans, A.; Boeve, A. Feeling bad or feeling good, does emotion affect your consumption of food? A meta-analysis of the experimental evidence. *Neurosci. Biobehav. Rev.* **2018**, *92*, 195–208. [CrossRef]
6. Kostyra, E.; Rambuszek, M.; Waszkiewicz-Robak, B.; Laskowski, W.; Blicharski, T.; Poławska, E. Consumer facial expression in relation to smoked ham with the use of face reading technology. The methodological aspects and informative value of research results. *Meat Sci.* **2016**, *119*, 22–31. [CrossRef]
7. Viejo, C.G.; Fuentes, S.; Howell, K.; Torrico, D.D.; Dunshea, F.R. Integration of non-invasive biometrics with sensory analysis techniques to assess acceptability of beer by consumers. *Physiol. Behav.* **2019**, *200*, 139–147. [CrossRef]
8. He, W.; Boesveldt, S.; de Graaf, C.; de Wijk, R.A. The relation between continuous and discrete emotional responses to food odors with facial expressions and non-verbal reports. *Food Q. Prefer.* **2016**, *48*, 130–137. [CrossRef]
9. Motoki, K.; Saito, T.; Nouchi, R.; Kawashima, R.; Sugiura, M. Tastiness but not healthfulness captures automatic visual attention: Preliminary evidence from an eye-tracking study. *Food Q. Prefer.* **2018**, *64*, 148–153. [CrossRef]
10. Leitch, K.; Duncan, S.; O'keefe, S.; Rudd, R.; Gallagher, D. Characterizing consumer emotional response to sweeteners using an emotion terminology questionnaire and facial expression analysis. *Food Res. Int.* **2015**, *76*, 283–292. [CrossRef]
11. Danner, L.; Haindl, S.; Joechl, M.; Duerrschmid, K. Facial expressions and autonomous nervous system responses elicited by tasting different juices. *Food Res. Int.* **2014**, *64*, 81–90. [CrossRef]
12. Crist, C.; Duncan, S.; Arnade, E.; Leitch, K.; O'Keefe, S.; Gallagher, D. Automated facial expression analysis for emotional responsivity using an aqueous bitter model. *Food Q. Prefer.* **2018**, *68*, 349–359. [CrossRef]
13. Den Uyl, M.; Van Kuilenburg, H. The FaceReader: Online facial expression recognition. In *Proceedings of Measuring Behavior*; Citeseer: University Park, PA, USA, 2005; Volume 30, pp. 589–590.
14. Kuhn, S.; Strelow, E.; Gallinat, J. Multiple "buy buttons" in the brain: Forecasting chocolate sales at point-of-sale based on functional brain activation using fMRI. *NeuroImage* **2016**, *136*, 122–128. [CrossRef]
15. Motoki, K.; Suzuki, S. Extrinsic factors underlying food valuation in the human brain. *PsyArXiv* **2020**, **3**, 1–18.
16. Li, S.; Deng, W. Deep facial expression recognition: A survey. *arXiv* **2018**, arXiv:1804.08348.
17. Ekman, P.; Friesen, W.V. Constants across cultures in the face and emotion. *J. Personal. Soc. Psychol.* **1971**, *17*, 124. [CrossRef]
18. Martinez, B.; Valstar, M.F.; Jiang, B.; Pantic, M. Automatic analysis of facial actions: A survey. *IEEE Trans. Affect. Comput.* **2017**, *10*, 325–347. [CrossRef]
19. Cai, J.; Meng, Z.; Khan, A.S.; Li, Z.; O'Reilly, J.; Tong, Y. Island loss for learning discriminative features in facial expression recognition. In Proceedings of the 2018 13th IEEE International Conference on Automatic Face & Gesture Recognition (FG 2018), Xi'an, China, 15–19 May 2018; IEEE: Piscataway, NJ, USA, 2018; pp. 302–309.
20. Zhao, J.; Mao, X.; Zhang, J. Learning deep facial expression features from image and optical flow sequences using 3D CNN. *Vis. Comput.* **2018**, *34*, 1461–1475. [CrossRef]
21. Li, Y.; Zeng, J.; Shan, S.; Chen, X. Occlusion aware facial expression recognition using cnn with attention mechanism. *IEEE Trans. Image Process.* **2018**, *28*, 2439–2450. [CrossRef]
22. Wang, Y.; Li, Y.; Song, Y.; Rong, X. Facial Expression Recognition Based on Auxiliary Models. *Algorithms* **2019**, *12*, 227. [CrossRef]
23. Liong, S.T.; Gan, Y.; See, J.; Khor, H.Q.; Huang, Y.C. Shallow triple stream three-dimensional cnn (ststnet) for micro-expression recognition. In Proceedings of the 2019 14th IEEE International Conference on Automatic Face & Gesture Recognition (FG 2019), Lille, France, 14–18 May 2019; IEEE: Piscataway, NJ, USA, 2019; pp. 1–5.
24. Gunaratne, T.M.; Fuentes, S.; Gunaratne, N.M.; Torrico, D.D.; Gonzalez Viejo, C.; Dunshea, F.R. Physiological responses to basic tastes for sensory evaluation of chocolate using biometric techniques. *Foods* **2019**, *8*, 243. [CrossRef]

25. Mahieu, B.; Visalli, M.; Schlich, P.; Thomas, A. Eating chocolate, smelling perfume or watching video advertisement: Does it make any difference on emotional states measured at home using facial expressions? *Food Q. Prefer.* **2019**, *77*, 102–108. [CrossRef]
26. Samant, S.S.; Seo, H.S. Using both emotional responses and sensory attribute intensities to predict consumer liking and preference toward vegetable juice products. *Food Q. Prefer.* **2019**, *73*, 75–85. [CrossRef]
27. Lagast, S.; Gellynck, X.; Schouteten, J.; De Herdt, V.; De Steur, H. Consumers' emotions elicited by food: A systematic review of explicit and implicit methods. *Trends Food Sci. Technol.* **2017**, *69*, 172–189. [CrossRef]
28. Kreibig, S.D. Autonomic nervous system activity in emotion: A review. *Biol. Psychol.* **2010**, *84*, 394–421. [CrossRef] [PubMed]
29. Wendin, K.; Allesen-Holm, B.H.; Bredie, W.L. Do facial reactions add new dimensions to measuring sensory responses to basic tastes? *Food Q. Prefer.* **2011**, *22*, 346–354. [CrossRef]
30. He, W.; Boesveldt, S.; Delplanque, S.; de Graaf, C.; De Wijk, R.A. Sensory-specific satiety: Added insights from autonomic nervous system responses and facial expressions. *Physiol. Behav.* **2017**, *170*, 12–18. [CrossRef]
31. Beyts, C.; Chaya, C.; Dehrmann, F.; James, S.; Smart, K.; Hort, J. A comparison of self-reported emotional and implicit responses to aromas in beer. *Food Q. Prefer.* **2017**, *59*, 68–80. [CrossRef]
32. Le Goff, G.; Delarue, J. Non-verbal evaluation of acceptance of insect-based products using a simple and holistic analysis of facial expressions. *Food Q. Prefer.* **2017**, *56*, 285–293. [CrossRef]
33. Palm, G.; Glodek, M. Towards emotion recognition in human computer interaction. In *Neural Nets and Surroundings*; Springer: Berlin/Heidelberg, Germany, 2013; pp. 323–336.
34. Monkaresi, H.; Bosch, N.; Calvo, R.A.; D'Mello, S.K. Automated detection of engagement using video-based estimation of facial expressions and heart rate. *IEEE Trans. Affect. Comput.* **2016**, *8*, 15–28. [CrossRef]
35. Gurney, K. *An Introduction to Neural Networks*; CRC Press: Boca Raton, FL, USA, 2014.
36. Mollahosseini, A.; Hasani, B.; Mahoor, M.H. Affectnet: A database for facial expression, valence, and arousal computing in the wild. *IEEE Trans. Affect. Comput.* **2017**, *10*, 18–31. [CrossRef]
37. Lucey, P.; Cohn, J.F.; Kanade, T.; Saragih, J.; Ambadar, Z.; Matthews, I. The extended cohn-kanade dataset (ck+): A complete dataset for action unit and emotion-specified expression. In Proceedings of the 2010 IEEE Computer Society Conference on Computer Vision and Pattern Recognition-Workshops, San Francisco, CA, USA, 13–18 June 2010; IEEE: Piscataway, NJ, USA, 2010; pp. 94–101.
38. Dalal, N.; Triggs, B. Histograms of oriented gradients for human detection. In Proceedings of the 2005 IEEE Computer Society Conference on Computer Vision and Pattern Recognition (CVPR'05), San Diego, CA, USA, 20–25 June 2005.
39. Kazemi, V.; Sullivan, J. One millisecond face alignment with an ensemble of regression trees. In Proceedings of the IEEE Conference on Computer Vision and Pattern Recognition, Columbus, OH, USA, 23–28 June 2014; pp. 1867–1874.
40. Zuiderveld, K. *Contrast Limited Adaptive Histogram Equalization*; Graphics Gems IV; Academic Press Professional, Inc.: London, UK, 1994; pp. 474–485.
41. King, D.E. Dlib-ml: A machine learning toolkit. *J. Mach. Learn. Res.* **2009**, *10*, 1755–1758.
42. Bradski, G. The OpenCV Library. *Dr. Dobb's J. Softw. Tools* **2000**, *25*, 120-125.
43. Chollet, F. Keras. 2015. Available online: https://github.com/fchollet/keras (accessed on 28 May 2020).
44. Abadi, M.; Barham, P.; Chen, J.; Chen, Z.; Davis, A.; Dean, J.; Devin, M.; Ghemawat, S.; Irving, G.; Isard, M.; et al. Tensorflow: A system for large-scale machine learning. In Proceedings of the 12th USENIX Symposium on Operating Systems Design and Implementation (OSDI 16), Savannah, GA, USA, 2–4 November 2016; pp. 265–283.
45. Breiman, L. Random forests. *Mach. Learn.* **2001**, *45*, 5–32. [CrossRef]
46. Rodger, H.; Vizioli, L.; Ouyang, X.; Caldara, R. Mapping the development of facial expression recognition. *Dev. Sci.* **2015**, *18*, 926–939. [CrossRef] [PubMed]
47. Calvo, M.G.; Nummenmaa, L. Perceptual and affective mechanisms in facial expression recognition: An integrative review. *Cogn. Emot.* **2016**, *30*, 1081–1106. [CrossRef]
48. Álvarez, V.M.; Sánchez, C.N.; Gutiérrez, S.; Domínguez-Soberanes, J.; Velázquez, R. Facial emotion recognition: A comparison of different landmark-based classifiers. In Proceedings of the 2018 International Conference on Research in Intelligent and Computing in Engineering (RICE), San Salvador, El Salvador, 22–24 August 2018; IEEE: Piscataway, NJ, USA, 2018; pp. 1–4.

49. Álvarez, V.M.; Domínguez-Soberanes, J.; Sánchez, C.N.; Gutiérrez, S.; López, B.; Quiroz, R.; Mendoza, D.E.; Buendía, H.E.; Velázquez, R. Consumer acceptances through facial expressions of encapsulated flavors based on a nanotechnology approach. In Proceedings of the 2018 Nanotechnology for Instrumentation and Measurement, Mexico City, Mexico, 7–8 November 2018; IEEE: Piscataway, NJ, USA, 2018; pp. 1–5.
50. Du, Y.; Zhang, F.; Wang, Y.; Bi, T.; Qiu, J. Perceptual learning of facial expressions. *Vis. Res.* **2016**, *128*, 19–29. [CrossRef]
51. Armstrong, J.E.; Laing, D.G.; Jinks, A.L. Taste-Elicited Activity in Facial Muscle Regions in 5–8-Week-Old Infants. *Chem. Sens.* **2017**, *42*, 443–453. [CrossRef]
52. De Wijk, R.A.; He, W.; Mensink, M.G.; Verhoeven, R.H.; de Graaf, C. ANS responses and facial expressions differentiate between the taste of commercial breakfast drinks. *PLoS ONE* **2014**, *9*, e93823. [CrossRef]

MDPI
St. Alban-Anlage 66
4052 Basel
Switzerland
Tel. +41 61 683 77 34
Fax +41 61 302 89 18
www.mdpi.com

Foods Editorial Office
E-mail: foods@mdpi.com
www.mdpi.com/journal/foods

www.ingramcontent.com/pod-product-compliance
Lightning Source LLC
LaVergne TN
LVHW070624170726
843515LV00004B/172

* 9 7 8 3 0 3 6 5 4 0 8 0 1 *